Developing Economies of the Third World

Outstanding Studies of
Economic Development in
Latin America and
the Pacific Rim

edited by
Stuart Bruchey

A Garland Series

CAPITAL LABOR SUBSTITUTABILITY IN MALAYSIAN MANUFACTURING

MAISOM ABDULLAH

GARLAND PUBLISHING, INC.
NEW YORK & LONDON
1991

Copyright © 1991 by Maisom Abdullah
All Rights Reserved

LIBRARY OF CONGRESS CATALOGING-IN-PUBLICATION DATA

Abdullah, Maisom.
Capital labor substitutability in Malaysian manufacturing/ Maisom Abdullah.
p. cm.—(Developing economies of the Third World)
Includes bibliographical references.
ISBN 0-8153-0627-X (alk. paper)
1. Capital productivity—Malaysia—Mathematical models.
2. Manufacturing processes—Malaysia—Automation—Mathematical models. 3. Malaysia—Manufactures—Employees—Supply and demand—Mathematical models.
4. Malaysia—Manufactures—Employees—Effect of technological innovations on mathematical models.
I. Title. II. Series.
HC445.5.Z9C36   1991
331.12'97'09595—dc20
91-25120

Designed by Lisa Broderick

Printed on acid-free, 250-year-life paper.
MANUFACTURED IN THE UNITED STATES OF AMERICA

PREFACE

Post-independent industrial development programmes in Malaysia have been assigned a central role in Malaysia's overall economic diversification strategy. Industrialisation is seen as the new source of rapid economic growth and employment opportunities, as well as a vehicle for greater Malay participation in economic activities in the country. It is also regarded as the agent that will bring about structural changes within the economy and is eventually expected to establish greater economic stability and a more equitable distribution of income and wealth.

Since its inception, the industrial policy has passed through four distinctive phases. The first was one of import substitution, which lasted from 1957 until 1968; the second was selective export-led industrialisation from 1968 until the early 1980's, the third phase from 1983 to 1986 was a period of divestment, retrenchment and general recession. The current phase which began in 1986 is a liberalised export-led industrialisation. The Industrial Master Plan, 1986-1995 identified 12 industrial sectors including 7 resourcebased industries for special expansionary treatment in the 1990's.

The early performance of the manufacturing sector was impressive. The rate of manufacturing output growth was 11.5 percent per annum between 1960 and 1970, and 12.5 percent per annum between 1971 and 1980. Although the recession years witnessed a slower growth of 5.0 percent per annum, manufacturing sector bounced back after 1986. The performance of exports and investments in the manufacturing sector was equally impressive. By 1987, the manufacturing sector contributed 22.4 percent of GDP, thus surpassing the contributions of agriculture, forestry and fishing. Since 1986, manufactured exports become the single largest foreign exchange earner.

Notwithstanding the impressive growth rates of output and exports and despite large amounts of investments, the growth rate of employment in the manufacturing sector have been less impressive. Growth rates of employment lagged behind that of output, investment and exports both in the manufacturing sector and in the economy as a whole.

In recent years, a number of literature has emerged attempting to explain this poor employment performance in the Malaysian economy. Many observers have felt that modern manufacturing techniques may not permit much substitution between labor and capital, and consequently that the ability of the manufacturing sector to absorb labor is quite limited. Others have pointed out that imperfections in the factor markets may account for the relatively slow growth of manufacturing employment. Firms may have been encouraged to adopt capital-intensive techniques by macro and foreign trade policies that in fact subsidize the importation of machinery. Furthermore, capital subsidies in the form of low interest rates and tax incentive, intended to attract investments from abroad, may have also the unwanted effects of stimulating capital-intensive production and encouraging excess capacity. Wage rates in the manufacturing sector have been increasing. Manufacturing wages tend to be higher than the supply price of labor from the traditional sector (e.g. rubber estates) by legislation on fringe benefits, shift differentials, and trade union pressures. Furthermore, exchange rates tend to be overvalued. It is likely that these factor market imperfections have affected the rate of labor absorption in the manufacturing sector. The central issue involved here concerns the degree of substitutability or the elasticity of substitution between capital and labor (σ.)

The focus of this work is the sectoral elasticities of substitution of Malaysian manufacturing industries. The interest in sectoral elasticities of substitution can be attributed to a number of reasons. Firstly, it is a continuing effort to explain the poor employment performance in various manufacturing industries. Secondly, the size of such elasticities is an important component of the degree of static flexibility of the different industrial sectors in response to external changes. Thirdly, the assumption of fixed proportion or zero elasticity underlies the use of input-output planning models used in Malaysia. Negation of this assumption would imply erroneous planning targets and policy adjustments. Finally, the higher the elasticity of substitution between capital and labor, the greater is the possible rate of growth of product, because the relatively fast growing primary factor can be substituted more easily for the relatively slow growing factor.

Elasticity of substitution between capital and labor is an important determinant of labor absorption. In a static analysis, low elasticity of substitution implies that the changes in capital-labor ratios due to changes in the marginal rate of technical substitution between the inputs will be small. As such, reduction in wage-rental ratios will result in negligible effect on labor absorption. In a more dynamic sense, labor absorption reflects the product mix, the choice of techniques and the scale of output, among other determinants on the demand side. As such, how much labor would be absorbed in the manufacturing sector in response to a given policy change would depend on changes in the scale of production as well as movements along given isoquants.

Most of the efforts to measure the degree of substitutability between capital and labor involve measurements in a CES production function with assumptions of perfect competition in both commodity and factor markets. The large number of econometric studies on possibilities for capital-labor substitution in the manufacturing sector in both developed and developing countries in general indicate some disagreements over the value of the elasticity of substitution; studies based on cross-sectional data provide estimates which are quite close to unity, but time-series studies generally report lower estimates. Furthermore, estimates of (σ) seem to vary systematically with the choice of functional form: regressions based on the marginal product of capital relation generally produce lower estimates than the regression based on the marginal product of labor relation. A variety of hypothesis have been advanced to explain the diversity of results, including cyclical changes in the utilisation of factors and random measurement errors, systematic variation of input price with product price, embodied and disembodied technical change and problems in the measurement of inputs, simultaneous equation bias, serial correlation and lagged adjustment. Empirical studies by Griliches (1967) and Lucas (1969) which attempt to take account of the difficiencies have produced unsatisfactry results.

In this study, econometric analysis of capital-labor substitutability is conducted for the Malaysian Manufacturing sector for the period 1963-1984. Estimates of the elasticity of substitution are computed using the Constant Elasticity of Substitution (CES) production function approach, the translog cost approach and the CES-translog cost approach. Comparison of alternative estimates of the elasticity of substitution between capital and labor are also made.

The statistical analysis shows that most of the Malaysian industries at the 5-digit industrial classification level exhibit relatively low elasticities of substitution between capital and labor. This provides a clearer understanding of the relatively low labor absorption in the Malaysian manufacturing sector. Other factors which may be responsible for the choice of inappropriate factor proportions including biases in government policy as well as private decisions by firms are also discussed.

In this study, attempts are made to reconcile the estimates of the elasticity of substitution in terms of simultaneous equation biases, as well as biases due to cyclical trends, under-utilisation and serial correlation. Nonetheless, they are still subjected to some other biases including random measurement errors, embodied and disembodied technical changes and lagged adjustment. Taking account of these remaining biases, we interpret that the actual elasticity estimates to be lower than the reported estimates. The time-series data used in this study actually sample a dynamic adjustment process due to a combination of factors such as changes in relative prices, technical changes and external shocks. The time-series estimates tend to be biased upwards because of simultaneity between the inputs and their prices and mis-specification of adjustment lags between the inputs and output and the dominance of cyclical conditions. While we are able to attempt to correct biases due to simultaneity and cyclical effects, the other remaining problems have biased the estimates of (σ) upwards.

Nerlove's (1967) and Morawetz's (1973) criticisms of econometric measurements of the elasticity of substitution was mainly due to their findings that in American industries 'even slight variations in the period or concepts used can produce different estimates of (σ). In this study however, it is interesting to note the similarities in the range and the generally low elasticity of substitution in the Malaysian manufacturing sector, either through the CES or the translog cost function approaches. It is however not possible to identify industries with consistently high or low elasticities and industry rankings too tend to be quite unstable. As such, it is not possible to interprete with confidence the point estimates of each industry.

Future studies on labor absorption in the manufacturing sector would require the development of a multi-commodity model of the industrial sector, incorporating own and cross-price elasticities, from which output supply and input demand relationships would be generated. Then output demand and input supply relations would have to be modelled, possibly exogeneously. Solving the model under different policy regimes would show how industrial employment would respond to given policy changes.

In undertaking this study, I am mostly indebted to Professor Lehman B Fletcher, Department of Economics, Iowa State University for this unwavering support throughout my Ph.D program. Professor Fletcher has also taught me the truest meaning of kindness. My undying gratitude is due to my husband, Ridzwan Halim for his example and love throughout the period of study. Finally, I would like to record my appreciation to Puan Noraini Suratman for typing the manuscript.

MAISOM ABDULLAH

Department of Economics
Universiti Pertanian Malaysia

March 1991

TABLE OF CONTENTS

	Page
CHAPTER 1. INTRODUCTION	1
The Concept and Significance of the Elasticity of Substitution (σ)	
Objectives of the Study	7
CHAPTER II. REVIEW OF THE LITERATURE	9
Anecdotal Evidence	9
Engineering or Process Analysis Studies	10
Econometric Investigation	11
Applications of Duality Theory	13
CHAPTER III. CONSTANT ELASTICITIES OF SUBSTITUTION (CES) PRODUCTION FUNCTION APPROACH	17
Methodology	17
The theoretical and empirical model of the elasticity of substitution under perfect competition.	18
Data, estimation and Discussion of Results of CES Production Function of the Malaysian Manufacturing sector	21
Sources of data and measurement of variables	21
Estimation of the elasticity of substitution and technical change under perfect competition	23
Discussion of empirical results	26
Elasticity of substitution and technical change under perfect competition, 1963 - 1984	34
Comparison of the CES estimates to other estimates of the elasticity of substitution.	44

CHAPTER IV. CAPITAL-LABOR SUBSTITUTION IN MALAYSIAN MANUFACTURING USING TRANSLOG COST FUNCTION APPROACH   49

Methodology   49

The theoretical and empirical model of the translog cost function   49

Measurement of technical change via the translog cost approach   55

Data, Estimation and Results of The Translog Cost Function of the Malaysian Manufacturing Industries   56

Sources of data and measurement of variables   56

Estimation of the two-input translog function   58

Discussion of empirical results   64

Elasticity of substitution without technical change   64

Elasticity of substitution with technical change   81

The theoretical and empirical model of the CES-translog cost function   81

CHAPTER V. CHOICE OF APPROPRIATE TECHNIQUE AND EMPLOYMENT GENERATION   87

Assessment of Technical Substitution Possibilities   87

Degree of product differentiation   89

Depth of transformation   89

Ease of mechanisation   90

Skill constraint   90

Factors Responsible for the Choice of Inappropriate
Techniques 90

    The influence of government policy biasses 91

    Biasses in private decision making 92

    "Engineering" versus "economic" man 92

    Costs and risks of technology search 93

    Position in international trade 93

    Role of multinational corporation 93

    Differences in labor productivity 94

CHAPTER VI. CONCLUSIONS AND POLICY IMPLICATION 97

    Directions for Further Research 99

BIBLIOGRAPHY 103

APPENDIX 109

LIST OF TABLES

Page

Table 1.1.	Comparison of growth rates of output, net exports and employment in Malaysia manufacturing sector, 1971-1987	1
Table 3.1.	Statistical performance of alternative models of the CES production function in Malaysian manufacturing sector, 1963-1984	28
Table 3.2.	Time series estimates of the elasticity of substitution of Malaysian manufacturing industries, 1963-1984	35
Table 3.3.	Ranking of elasticity of substitution in Malaysian manufacturing industries, 1963-1984	37
Table 3.4.	Tests of significance of the elasticity of substitution of Malaysian manufacturing industries	40
Table 3.5.	Time series estimates of the elasticity of substitution and the rate of technical progress in Malaysian manufacturing industries, 1963-1984	42
Table 3.6.	Comparison of alternative estimates of elasticity of substitution based and CES production function approach	45
Table 4.1.	Statistical performance of alternative models of the unrestricted translog cost function estimation of Malaysian manufacturing industries, 1969-1984	65
Table 4.2.	Unrestricted SUR parameter estimates of capital-labor translog cost function estimation of Malaysian manufacturing industries, 1969-1984	71
Table 4.3.	Restricted estimates of capital-labor translog cost function for Malaysian manufacturing sector, 1969-1984	77
Table 4.4.	Elasticity of substitution in Malaysian manufacturing industries: a translog cost function approach	79

LIST OF FIGURES

		Page
Figure 1.1.	σ is small	4
Figure 1.2.	σ is high	5
Figure 1.3.	$\sigma = 0$	6

CHAPTER 1
INTRODUCTION

One of the most important issues facing the Malaysian Government today is mounting population and unemployment pressures. Malaysia's population has been growing at 2.6 percent per annum, from a population of 10.9 million in 1970 to 13.9 million in 1980 and 16.5 million in 1987. The labor force however grew faster than population at 3.5 percent per annum and has resulted in an increasing rate of unemployment from around 3 percent in 1970 to 5.7 percent in 1980 and 8.7 percent in 1987. The rate of unemployment is expected to increase further to 10.1 percent by 1990 (Government of Malaysia, 1987).

Since the New Economic Policy was formulated in 1970, the manufacturing sector is envisaged to play the pivotal role in creating new employment opportunities. Due to vigorous efforts to develop the industrial sector, manufacturing has become one of the fastest growing activities in the economy. However, there is a growing body of evidence to show that the growth of industrial employment is lagging behind the growth of output and exports. [Hoffman & Tan, 1980; Young, 1980; Jomo, 1985].

As shown in Table 1.1, despite fairly rapid rates of growth of output and exports, the rate of employment growth has persistently been relatively low. For example, in the boom years of late 1970's the rate of growth of output and net exports reached double digit figures of 13 and 20 percent respectively. The rate of growth of employment for the same period was only 3.7 percent.

Table 1.1 Comparison of growth rates of output, net exports and employment in Malaysian manufacturing sector 1971 - 1987.

	Output	Net Exports	Employment
1971 - 73	16.1	18.8	9.5
1976 - 80	13.0	20.0	3.7
1981 - 83	4.9	17.1	2.2
1986 - 87	12.5	14.3	6.2

(Source: Government of Malaysia, Development Plan 1974, p. 142 - 47, Development Plan 1984, p. 249)

In recent years, a voluminous literature has emerged attempting to explain this poor employment performance. Most of the research come to the conclusion that the unemployment problem is multi-faceted and related to the structure of the economy. The consensus is that the problem of generating more employment opportunities involves a substantial redefinition of appropriate development strategies involving these fundamental and inter-related questions: Is there a conflict between increasing employment and accelerating growth? Which goods should be produced (output-mix problem)? How should capital and labor be used to produce final goods (choice of technique problem)?

The last question above has emerged as an important area for research and for policy in LDC's. Many observers have felt that modern manufacturing techniques do not permit much substitution between labor and capital, and consequently that the ability of the manufacturing sector to absorb labor is quite limited. Others have pointed out, however, that imperfections in the factor markets might account for the relatively slow growth of manufacturing employment. Firms may have been encouraged to adopt capital-intensive techniques by macro and foreign trade policies which in fact subsidize the importation of machinery. Furthermore, capital subsidies in the form of low interest rates and tax incentives, intended to attract investments from abroad, may have also the unwanted effects of stimulating capital-intensive production and encouraging excess capacity. Wage rates in the manufacturing sector have also been pushed up above the supply price of labor from the traditional sector by legislation on fringe benefits, shift differentials, penalties on firing,and trade union pressures. It is likely that these factor market imperfections have affected the rate of labor absorption in the manufacturing sector. The central issue involved here concerns the degree of substitutability or the elasticity of substitution between labor and capital.

The Concept and Significance of the Elasticity of Substitution (σ)

The elasticity of substitution (σ) is a technical parameter characterising a production function [Nadiri, 1970]. It is a measure of the ease with which any two inputs such as capital and labor can be substituted for each other. The elasticity of substitution is defined as the ratio of the proportionate change in factor proportions to the proportionate change in the slope of the isoquant or the marginal rate of technical substitution (MRTS). With the assumption of cost-minimising behavior of firms, the marginal rate of technical substitution is equal to the ratio of factor prices. Thus, for a general production function of the form,

$Q = f(K,L)$ where K is capital, L is labor and Q is output, cost minimisation is shown by the lagrangian function,

$$\min l = WL + RK + [f(K,L) - Q]$$

Taking first partial derivatives and setting them equal to zero gives

$$\frac{\delta l}{\delta K} = R + f_K = 0$$

and the marginal product of labor is

$$\frac{\delta l}{\delta L} = W + f_L = 0$$

The marginal rate of technical substitution between capital and labor ($MRTS_{K,L}$) is equal to the wage-rental ratio at the optimal point.

$$MRTS_{K,L} = \frac{f_L}{f_K} = \frac{W}{R}$$

The elasticity of substitution between capital and labor can thus be defined as

$$\sigma = \frac{\% \text{ change in } (K/L)}{\% \text{ change in } MRTS_{K,L}} = \frac{d \log (K/L)}{d \log (W/R)}$$

i.e, the proportionate change in factor proportions to a proportionate change in relative factor prices. Thus, for example, if the elasticity of substitution is 0.1, then a 10 percent change in relative factor prices will bring about a one percent change in factor proportions (0.1 x 10). Graphically, when the elasticity of substitution is low a relatively large change in the wage-rental ratio from $(W/R)^A$ to $(W/R)^B$ will result in only a small change in the capital-labor ratio. Demand for labor increases from L^* to L_1 while demand for capital decreases for K^* to K_1 (Figure 1.1)

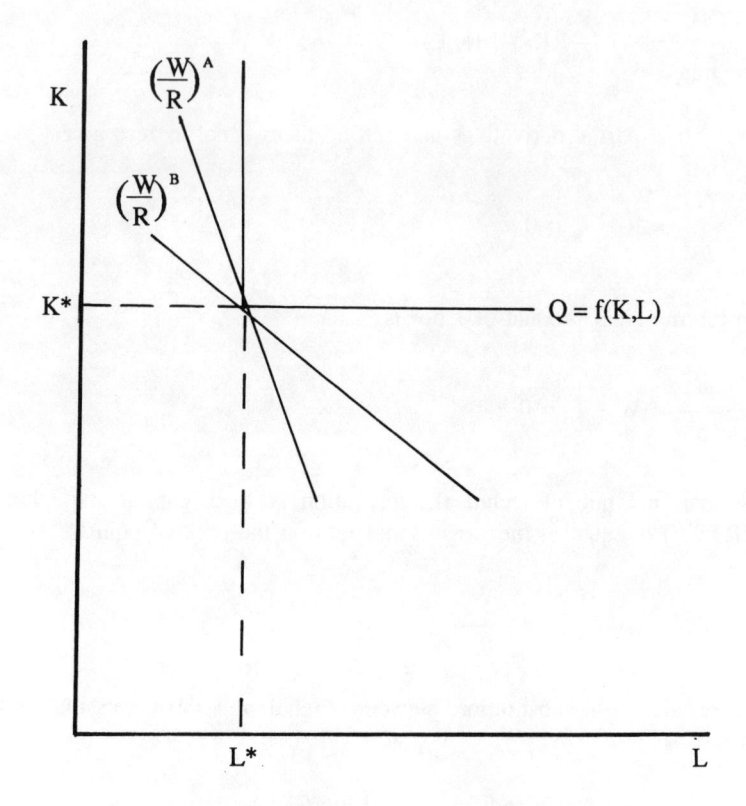

Figure 1.1: σ is small

If on the other hand, the elasticity of substitution is 5.0, then a 10 percent change in relative factor prices will bring about a 50 percent change in factor proportions. The high elasticity of substitution between capital and labor is depicted by Figure (1.2). An equal change in wage-rental ratio from $(W/R)^A$ to $(W/R)^B$ will cause a larger change of capital-labor ratio and result in great increase in demand for labor from L^* to L_2. In the case of fixed proportions, i.e $\sigma = 0$, then changes in the relative factor prices will not change factor proportins at all, as shown in Figure 1.3.

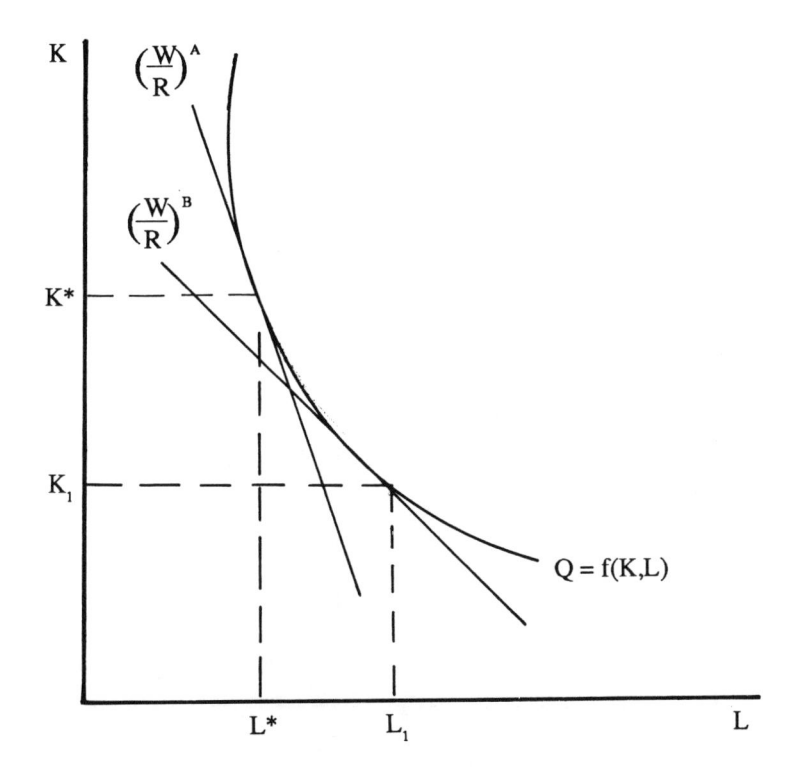

Figure 1.2: σ is high

The elasticity of substitution ranges from zero to infinity. In the two-factor production function, a positive elasticity of substitution indicates that efficient factor substitutability is possible. Theoretically, when the elasticity of substitution is negative, it implies complementarity, In the two-factor case considered here, it is not possible to maintain production and decrease the use of both inputs as would be implied by complementarity when wages are increased. For the neo-classical two-factor production function, negative elasticities therefore imply inefficiencies. Only when more than two factors are considered, it is possible for complemnetarity to exist. However, at least two of the n factors must be substitutes.

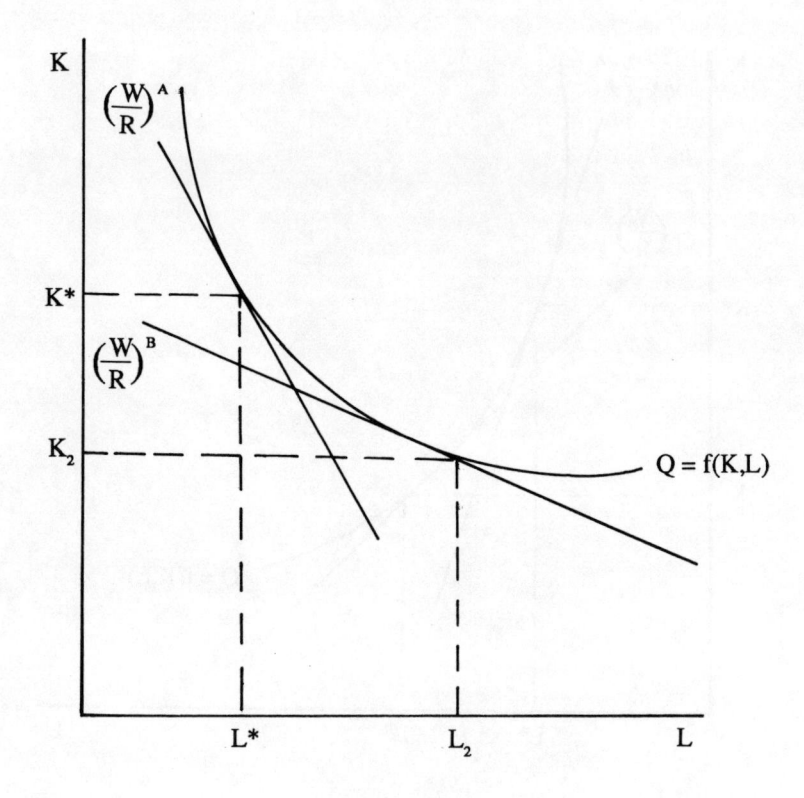

6 *Substitutability in Malaysian Manufacturing*

Figure 1.3: $\sigma = 0$

Conceptually, the elasticity of substitution has a number of important policy implications. The elasticity of substitution is a measure of the ease with which any two inputs such as capital and labor can be substituted for each other. If substitution between factors is relatively easy, then competitive firms will be induced to absorb increased employment by a relatively small reduction in wage-rental ratio. It follows that knowledge of the values of the elasticity of substitution in industrial sectors and subsectors are not only useful for policy-makers for changing the market signals to ensure greater labor absorption, but is also useful in identifying

appropriateness of techniques that are being used. If the value of the elasticity of substitution is found to be near zero in a labor-surplus economy, then it is imperative to adopt a new technology with greater substitutability. Furthermore, encouraging labor intensive industries through factor-price manipulations would be meaningless and costly if in fact factor substitution is low. It is therefore important to discuss the following questions concerning Malaysian industrial production;
1) Are elasticities of substitution between capital and labor significantly different from zero?
2) What conclusions can be drawn about the influence of policies from the experience of Malaysia's industrialisation process?
3) What are the policy implications that can be derived from the results of this study?

Objectives of the Study

The principal objective of this dissertation is to provide an econometric analysis of the capital - labor substitution possibilities in the Malaysian manufacturing industries. Specifically, the objectives are:
1) to provide the theory, specification and the estimation procedures for the elasticity of substitution between labor and capital in Malaysian manufacturing industries using three different approaches, i.e.
The Constant Elasticity of Substitution (CES) Production Function
The Translog Cost Function
The CES - Translog Cost Function
2) to discuss comparability of alternative estimates of the elasticity of substitution
3) to evaluate and draw policy implications from the empirical findings

The present study is organised as follows:
Chapter II presents a brief review of the relevant literature on the alternative approaches to estimating the elasticity of substitution. Chapter III presents the theoretical model, the appropriate estimation procedure and the interpretation of results based on the CES production function. Chapter IV provides the theoretical models, the estimation and interpretation of results based on the translog cost function and the CES - translog cost function approaches. Chapter V provides a discussion of factors which may have influenced the choice of technique in a developing country such as Malaysia. Conclusions and some policy implications are discussed in the final chapter.

CHAPTER 11
REVIEW OF THE LITERATURE

Since the early 1970's the question of capital-labor substitution possibilities in industrial production in developing countries has received widespread attention by policy makers and researchers. While most researchers focus on the question of appropriate factor proportions, other studies apply engineering or process analysis. Still others provide reports which do not offer precisely quantified estimates of the efficient production frontier. This last group of studies is referred to as anecdotal evidence. This review of literature will focus on factor proportions and the various ways of improvement on the methodology to measure the elasticity of substitution between labor and capital.

Anecdotal Evidence

This category includes studies that discuss labor-capital substitution and other factors affecting technological choice but do not offer precisely quantified estimates of the frontier of efficient combinations. Since the 1960's anecdotal literature has grown tremendously. This form of literature offers useful insights into the problems of appropriate technology transfer to LDCs and point to the same conclusions as other studies. Some of this literature is reviewed below.

Stewart (1972) argued that multinational corporations are the major deterrent in appropriate choice of technology in LDCs, since they are tied to their capital-intensive technology in the developed countries. Efforts to change and adapt their technology to LDC conditions are costly and risky. They ignore the possibility of using local raw material inputs. And, even if they are considering adaptations, they frequently pay higher wages than local firms and hence use less labor per unit of output.

Schumacher (1973), who first introduced the concept of intermediate technology and was the founder of the Intermediate Technology Development Group (ITDG), believed that developing countries need the kind of technology which are cheap, and can be used without sophisticated technical or organizational skills and which are associated with small-scale enterprises.

Costa (1973) suggested that, in order to compensate for their scracity of capital and to provide jobs for rural populations, the developing countries should adopt labor-intensive methods. The labor-intensive programs need to be flexible and constantly adjustable to changes in manpower ability.

R.S. Eckaus (1977) pointed out that the criteria of appropriateness of technical decisions depends on the goals of development. These criteria concerned not only with income and quantities of outputs but also with the way output are produced and distributed. These criteria for appropriate technology include maximization of net national output and income, maximization of availability of consumption good, maximization of rate of economic growth, redistribution of income and wealth, regional development, promotion of political development and national political and social goals and improvement in the quality of life. Technological choices according to Eckaus are affected by resources and product markets, taxes, and policies. As such, it is very difficult to achieve the most appropriate technology.

Ranis (1979) defined appropriate technology relative to society's resources and goals. Ranis emphasized institutional and social factors as well as prices as important in determining the choice of appropriate technology.

Wells (1984) argued that firms in LDCs choose inappropriate technology because decisions are made by the 'engineering man' and monopolistic firms. Engineering man tends to choose highly advanced capital-intensive technologies in order to produce high quality products. Monopolistic firms tend to adopt capital-intensive technologies because they can pay for insurance coverage against risks.

Engineering or Process Analysis Studies

In these kinds of studies, researchers investigate individual manufacturing processes or individual products in order to determine appropriate technologies. The investigators usually use engineering or other technical information to determine the inputs necessary to produce a given volume of products. A principal part of the investigation is to see if there are alternative means of producing a product, that is, if more workers and fewer machines (or simpler or cheaper machines) can produce more cheaply than fewer workers and more machines. Published evidence exists for only a dozen products and processes a few of which are reviewed below. Such studies involve very high costs and are normally funded by international organizations.

Fong Chan Onn (1980) attempted to operationalise the analysis of appropriate technology based on Eckaus's criterion for evaluating appropriateness in reference to the development goals of Malaysia. Using detailed engineering and cost data for the bicycle manufacturing industry, he concludes that Raleigh technology is not appropriate technology. He suggests that the most appropriate technology for bicycle manufacturing in Malaysia is an improved version of Chinese bicycle technology (with improvements in finishing and quality control).

Hill (1983) modified the neoclassical approach to the choice of technology. Using detailed cost data on four alternative techniques of Indonesian weaving industry, he constructs a modified isoquant representing the capital and labor requirements for each technique. Hill then discussed the results and concluded that the Indonesian Government's policies (prohibition of the import of used machinery, exemption of the duty on imported machinery, an overvalued exchange rate) have encouraged the use of excessively capital-intensive techniques in the Indonesian weaving industry.

Timmer (1984) studied four alternative rice-milling processes and rice marketing systems in Indonesia. He found that the most capital-intensive technique required $65,000 investment per worker and the most labor-intensive technique requires only $700 per worker.

Pack (1984) analysed past choices of technology and present levels of productivity in the Philipines' cotton spinning and weaving industries. Detailed engineering and economic information were used to assess the costs of alternative technologies and to estimate the levels of productivity relative to international standards of best practice. He then concluded that high costs of production are due both to inappropriate technological choices and low productivity in the use of the technologies chosen.

Econometric Investigation

Since the seminal work of Arrow, Chenery, Minhas, and Solow (1961), a number of researchers have used LDC data to measure the degree of substitutability between capital and labor. The efforts here involved measurements of the elasticity of substitution using a CES production function involving capital and labor inputs. Since the CES production function is nonlinear and cannot be estimated through ordinary least-squares estimation techniques and since data on capital is frequently not available or not considered reliable, an indirect method is used. Normally, the logarithm of value-added per worker is regressed against the logarithm of the wage. The coefficient of the latter is an estimate of the elasticity of substitution. Some studies regress the output-capital ratio against the return of capital (both in logarithms) or regress the capital-labor ratio against the wage (both in logarithms) to provide alternative estimates. These estimates have been made for both cross-section and time-series data [Diwan, 1965; Hoffman & Tan, 1980].

Behrman (1972) used estimating equations based on CES and semi-Cobb-Douglas production functions to measure sectoral elasticities of substitutions between capital and labor in Chile. Using the Koyck-Cagan-Nerlove adjustment process, Behrman concluded that the degree of sectoral flexibility in response to factor price changes was limited in the Chilean case.

Lianos (1975) reported his estimation of the elasticity of the capital-labor substitution and the rate of technical change in the manufacturing sector of the Greek economy. Using the Cobb-Douglas production function with technical progress, four sets of substitution elasticity estimates were obtained; two time-series sets and two cross-section sets. He found that the elasticity of substitution for Greek manufacturing industries exceeded unity, and the annual rate of neutral technical change is $e^{0.05} = 1.05$.

O'Donnell and Swales (1979) generated the elasticity of substitution for UK manufacturing industries by using a variant of normal logarithmic transformation of CES production function. Using pooled data and Maximum Likelihood estimation procedures, their 'best' estimates of the substitution elasticity for UK manufacturing industries ranged between 0.4 and 1.6.

Claque (1969) modified the estimating equation to conform to a number of limiting assumptions for the Peruvian manufacturing sector. His estimating equation was a ratio of the Peruvian and the American situation which he regards as the standard measure for comparison. He concluded that underdeveloped countries (Peru) buy the vast majority of their machines and technology form high-wage countries (USA). Furthermore, the capital-labor ratio was lower in Peru because Peruvian workers, being less skilled, cannot handle machines as well as American workers.

More recent estimation of the elasticity of substitution using the CES production function approach were based on Korean manufacturing data (Jae Won Kim, 1984) and Singaporean manufacturing data (Toh Mun Heng, 1986). The CES production function approach had also been applied in studying substitution between production labor and other inputs in unionised and nonunionised American manufacturing sector (Freeman and Medoff, 1982).

This review of the literature on the estimation of elasticity of substitution based on the CES production function and its variants highlights two important issues, namely the statistical problems associated with cross-section and time-series studies of production functions and the choice of estimating equations. With respect to the first issue, time-series data usually reflect a dynamic adjustment due to a combination of factors such as changes in relative prices and external shocks, which are generally excluded in cross-section data. The time-series estimates are often biased because of simultaneity between the inputs and their prices (simultaneous equation bias) and mis-specification of adjustment lags between inputs and outputs, and the dominance of cyclical conditions – like under-utilization of capacity.

The cross-section results are also plagued by certain conceptual and estimation problems. In a competitive market there is no reason for relative prices to differ among production units. Any observed differences in intra-firm managerial ability and consequently are due to differences in skill or in the quality of inputs, then cross-section estimates of the elasticity of substitution will be biased towards unity [Guade, 1975].

Attempts which have been made to remove the restrictive features of the CES production function have taken two forms; amendments to the CES production function and indirect estimation of the production function by formulating relevant cost functions (quadratic, leontief, transcendental, Box-Cox) and other flexible functional forms [Diewert, 1971, 1973; Christensen, Jorgenson and Lau, 1973].

Lu and Fletcher (1968) provide an amendment to the CES production function. This is known as the Variable Elasticity of Substitution (VES) production function. Using this function, the invariance of elasticity of substitution (σ) to capital - intensity is tested by fitting the relation:

In (Q/L) = In a + b In (w/p) + c In (K/L) + e

and the elasticity of substitution for this function is

$$\sigma = \frac{1}{1 + \rho - m_p/s_K}$$

where ρ = [(1 - b)/b]

m_p = [c/1 - b)

s_K = the share of capital

Applications of Duality Theory

In recent years, the application of the duality theory has become increasingly popular among economists in applied economic analysis. This is because the methodology not only allows researchers greater flexibility in the specification of factor demand and output supply equation, but also permits a very close relationship between theory and practices. Furthermore, there are a number of advantages in estimating the elasticity of substitution by the cost function approach. Firstly, the multicollinearity problems which are inherent in the production function approach due to high correlation between inputs are less severe, because prices are formed in separate factor markets. Secondly, the elasticity of substitution is linearly related to the estimated parameters and thus their econometric properties are well defined. Finally, no matter what the properties of the production function are, the dual cost function is always linearly homogeneous in the prices and, as a result, the estimating procedure is more general.

Several functional forms have been used to estimate the cost function. They are the Cobb-Douglas (CD), the Constant-Elasticities of Substitution (CES), the Generalized Leontief (GL), the Generalised Box-Cox (GBC), the Fourier Flexible Form and the Transcendental Logarithmic (translog) Form (TLC). The translog cost function approach is the most popular functional form that has been used to analyse factor substitution in manufacturing sector [Berndt and Christensen, 1973; Humphrey and Moroney, 1975; Halvorsen, 1977; Wills, 1979; Rushdi, 1982; Tsao Yuan, 1986]. The translog cost function approach has also been used to study energy substitution effects (Berndt and Wood, 1975; Fuss, 1977; Field, 1980; Vashist, 1985], factor substitution and technical change in agricultural sector [Biswanger, 1974); Lopez, 1980; Ray 1982] and other branches of economics such as natural resources (Halvorsen and Smith, 1986) and labor economics (Freeman and Medoff, 1982).

Berndt and Wood (1975) used a translog cost function to test for price elasticity and the elasticity of substitution of capital, labor, energy and raw material inputs in US manufacturing for 1947-1971. By using the iterative 3-stage least-square procedure (I3SLS) on time-series data, Berndt and Wood have been noted as pioneers in effectively providing statistical procedures which meet the translog restrictions of equality, symmetry, homogeneity, and concavity.

Griffin and Gregory (1976) attempted to improve Berndt and Wood's work by applying a similar translog methodology to a pooled international data for manufacturing. They concluded that their model provides a reasonable long-run alternative to the time-series model. Pooled data however require the use of weighted least-square procedure which may result in simultaneous-equation bias and specification error problems in the estimation of a translog cost-function.

The development of literature of the cost-function approach has taken two trends: (a) application of the translog cost approach and (b) further testing and development of improved functional forms and improved data-specifications. Vashist (1985) applied the translog cost function approach to estimate the substitution possibilities and price sensitivity of energy demand in Indian manufacturing for the period 1961-1970, following a similar procedure as Brendt and Wood. Tsao Yuan (1986) similarly applied the translog cost function to estimate factor substitutability in the Singaporean manufacturing industries.

Most recent literature proposed alternative ways to measure the Allen Elasticities of Substitution (AES). These include Pollack, Sickles and Wales (1984); Elbadawi, Gallant and Souza (1983) and Jae Wan Chung (1987).

Pollack, Sickles and Wales (1984), proposed and estimated a new single porduct cost function, called the CES-translog cost function. Like the translog, it is a flexibel functional form, but it is compatible with a wider range of substitution possibilities than either the CES or the translog. As the name suggests, the CES-translog includes both the CES production function and the translog cost function as special cases and thus permits nested testing using conventional statistical techniques. Using the same time-series data used by Berndt and Wood (1985), they estimated the new CES-translog function to the US manufacturing sector and concluded that it was significantly superior to the translog. The CES-translog function however must satisfy additional restrictions of regularity conditions globally which are not tested in this paper.

Elbadawi, Gallant and souza (1983) digressed from using the translog cost approach in their estimation of price and substitution elasticities. They explored the possibility of estimating the elasticities using the fourier flexible form with commonly used statistical methods; multivariate least-squares, maximum likelihood (MLE) and three-stage least squares (3SLS). They concluded that the elasticities can be estimated consistently without a priori knowledge of functional form provided the number of fitted parameters were chosen adaptively by observing the data. The number of fitted parameters must increase as the number of observations increase. However, credibility of their estimates have not been tested for negative-semi-definiteness of the fourier flexible form.

Jae Wan Chung (1987) estimated elasticities of substitution via a truncated, single translog cost-share equation. Jae attempted to make comparisons and to reconcile elasticity estimates using different methods of estimation, differing data and observations and under differing assumptions.

This section has summarised the available literature on the possibilities of capital-labor substitution in the manufacturing sector. Each subsection has presented the development of different aspects of evidence, encompassing anecdotal literature, engineering or process analysis studies and econometric analysis. The focus of this dissertation is the study of capital-labor subsitution possibilities in Malaysian manufacturing sector through different types of econometric analysis which will be dealt with in Chapters III and IV respectively.

CHAPTER III
CAPITAL-LABOR SUBSTITUTION IN MALAYSIAN MANUFACTURING USING CONSTANT ELASTICITY OF SUBSTITUTION (CES) PRODUCTION FUNCTION APPROACH

Methodology

The Arrow - Chenery - Minhas - Solow or the CES production function has been well received and extensively analysed since its introduction in 1961 [Arrow, et al, 1961], both theoretically and empirically.
The CES production function is given by

$$Y = \gamma [\, \delta K^{-\rho} + (1-\delta) L^{-\rho}\,]^{-\nu/\rho} \qquad (3:0)$$

where Y is output, K and L are factors of capital and labor, and γ, δ and ρ are constants denoting efficiency, distribution and substitution parameters. ν denotes the degree of homogeneity. With assumptions of perfect competition in both commodity and factor markets and constant returns to scale, the CES production function is

$$Y = \gamma [\, \delta K^{-\rho} + (1-\delta) L^{-\rho}\,]^{-1/\rho} \qquad (3:1)$$

The elasticity of substitution is derived from the marginal products of labor and capital respectively.

$$\frac{\delta Y}{\delta L} = \gamma (1-\delta) L^{-\rho -1} [\delta K^{-\rho} + (1-\delta) L^{-\rho}]^{-1/\rho -1} \qquad (3:2)$$

$$\frac{\delta Y}{\delta K} = \gamma (\delta) K^{-\rho -1} [\delta K^{-\rho} + (1-\delta) L^{-\rho}]^{-1/\rho -1} \qquad (3:3)$$

In competitive equilibrium, the marginal rate of technical substitution $MRTS_{K,L}$ equals the factor price ratio. Thus,

$$MRTS_{K,L} = \frac{(1-\delta)}{\delta} [\frac{K}{L}]^{1+\rho} = \frac{W}{R} \qquad (3:4)$$

18 Substitutability in Malaysian Manufacturing

The elasticity of substitution is defined as

$$\sigma = \frac{\% \text{ change in } (K/L)}{\% \text{ change in } MRTS_{KL}}$$

$$= \frac{d \ln (K/L)}{d \ln MRTS_{KL}}$$

From (3:4)

$$\ln MRTS_{KL} = \ln \left(\frac{1-\delta}{\delta}\right) + (1+\rho) \ln (K/L)$$

$$\ln(K/L) = \left(\frac{1}{1+\rho}\right) \ln MRTS_{KL} - \left(\frac{1}{1+\rho}\right) \ln \left[\frac{1-\delta}{\delta}\right]$$

$$\frac{d \ln (K/L)}{d \ln MRTS_{KL}} = \sigma = \left(\frac{1}{1+\rho}\right) \tag{3:5}$$

For $\sigma > 1$, an increase in the $MRTS_{KL} = \dfrac{W}{R}$ by 1% implies that the capital - labor ratio (K/L) will increase by more than 1%. This means a small increase in the wage - rental ratio will lead to a relatively large reduction in the demand for labor resulting in greater unemployment.

The theoretical and empirical model of the elasticity of substitution under perfect competition

A Statistical estimation of the elasticity of substitution cannot be derived directly from equation [3:1]. Furthermore, since data on capital stock in Malaysia is questionable, an indirect method following Ferguson (1965) is used to derive the estimating equations without using the capital stock data. Starting with the CES production function,

$$Y = \gamma [\delta K^{-\rho} + (1 - \delta) L^{-\rho}]^{-1/\rho} \tag{3:1}$$

Constant Elasticity of Substitution 19

Let $y = Y/L$

Then $y = \gamma \: [\delta \: (\frac{K}{L})^{-\rho} + (1-\delta)]^{-1/\rho}$ \hfill (3:6)

Raising both sides of equation (3:6) to the ρ, we obtain

$$y^{\rho} = \gamma^{\rho} \: [\delta (\frac{K}{L})^{-\rho} + (1-\delta)]^{-1} \hfill (3:7)$$

$$y^{\rho} . \gamma^{-\rho} = [\delta (\frac{K}{L})^{-\rho} + (1-\delta)]^{-1} \hfill (3:8)$$

Assuming competitive equilibrium, where the value of the marginal product of labour equals the wage rate, we have

$$\frac{\delta Y}{\delta L} = W = \gamma \: (1-\delta) L^{-\rho - 1} \: [\delta (\frac{K}{L})^{-\rho} + (1-\delta)]^{-1/\rho - 1} \hfill (3:9)$$

Substituting equation (3:6) into equation (3:9),

$$W = y(1-\delta) \: [\: \gamma \: (\frac{K}{L})^{-\rho} + (1-\delta)]^{-1} \hfill (3:10)$$

Then, we substitute equation (3:8) into (3:10) and we get

$$W = y(1-\delta) \: y^{\rho} . \gamma^{-\rho} \hfill (3:11)$$

$$= (1-\delta) \: y^{1+\rho} \gamma^{-\rho} \hfill (3:12)$$

20 Substitutability in Malaysian Manufacturing

Transforming,

$$y^{(1+\rho)} = W \gamma^\rho (1-\delta)^{-1} \qquad (3:13)$$

Taking the logarithm of equation (3:13), and dividing by $(1+\rho)$, and transforming, we get

$$\ln y = \left(\frac{1}{1+\rho}\right) \ln \gamma^\rho (1-\delta)^{-1} + \left(\frac{1}{1+\rho}\right) \ln W \qquad (3:14)$$

$$\ln y = \sigma_1 \ln \gamma^\rho (1-\delta)^{-1} + \sigma_2 \ln W \qquad (3:15)$$

In equation [3:15], the efficiency parameter is considered a constant. However, if we assume that the technology becomes more efficient through time, then the production function will be shifted upwards in a neutral way. To show this change in technology, the efficiency parameter can be written as

$$\gamma = e^{\lambda t} \qquad (3:16)$$

where λ indicates the rate of neutral technical progress.

Replacing the value of γ in equation (3:13) we have

$$y^{(1+\rho)} = W(e^{\lambda t})^\rho (1-\delta)^{-1} \qquad (3:17)$$

Proceeding the same way as before,

$$\ln y = \left(\frac{1}{1+\rho}\right) \ln W + \left(\frac{1}{1+\rho}\right) \lambda t - \left(\frac{1}{1+\rho}\right) \ln (1-\delta) \qquad (3:18)$$

Equation (3:18) can now be written as

$$\ln(Y/L) = a_1 + a_2 \ln W + a_3 t \qquad (3:19)$$

where

$$a_1 = -\left(\frac{1}{1+\rho}\right) \ln(1-\delta)$$

$$a_2 = \text{estimate of the elasticity of substitution} = \left(\frac{1}{1+\rho}\right)$$

$$a_3 = \left(\frac{\rho}{1+\rho}\right), \text{ which permits the estimation of } \lambda \text{ once we have estimated } \rho$$

$$\lambda = \left(\frac{a_3}{1-\sigma}\right) \text{ is the annual rate of technical progress}$$

[Note $\rho = \left[\dfrac{1}{\sigma} - 1\right]$]

$$\lambda = \left(\frac{a_3}{1-\sigma}\right)$$

Y/L = output per unit of labor

W = wages and salaries

Data, Estimation and Discussion of Results for CES Production Function.

This section consists of discussion of the data base, the operational definition of variables, the estimation procedure in the calculation of the elasticity of substitution and a discussion of results of the estimation of the CES production function in Malaysian manufacturing industries.

Sources of data and measurement of variables

The data required for the estimation of the CES production functions are reported in the Censuses/Surveys of Manufacturing Industries, West

Malaysia for the period 1963 to 1976, and the Industrial surveys of Malaysia for 1978 to 1984. Consistency of time series data was achieved for the 5-digit Malaysian Industrial classification by focussing only on West Malaysia. Since the Industrial surveys of Malaysia report revenues and expenditures, the value-added figures for 1982, 1983 and 1984 were obtained directly from the Manufacturing division, Department of Statistics, Malaysia. Furthermore, missing observations for 1977 and 1980 for all reported industries were estimated using the linear interpolation method by fitting the average of the two intervening years. An attempt is made to estimate the missing values using zero-order interpolation method, i.e. by fitting in the missing values for 1977 and 1980 by their forecasted values. This is carried out for 10 industries as a test of whether zero-order interpolation will produce better results. Since it made little difference to the results, the final estimation was based on the first method, whereby, given Y_t and Y_{t+2}

$$\text{then } T\hat{y} = \frac{Y_{t+2} - Y_t}{2}$$

$$\text{and } Y_{t+1} = Y_t + T\hat{y}$$

For each industry, the data required for estimation are as follows: value-added in current Malaysian Ringgit (VA), wages and salaries (W), labor (L), time (T), Consumer Price Index (CPI) and Industrial Production Index (IPI). Number of labor is calculated as full-time plus half of part-time workers. No attempt is made to construct the Divisia quality index of labor because the data on hours utilised per person and educational attainment of labor in each 5 - digit classification industry groups is not available in the Surveys/Censuses of Manufacturing Industries of Malaysia. Estimated adjustments may result in inconsistent time series. The construction of the Divisia Quality index of labor is based on the stock of labor as measured by persons engaged, adjusted for effective hours per person, and changes in the composition of labor by educational attainment, age distribution and sex composition of the labor force. [Christensen and Jorgenson (1970), Feldstein (1967)].

The concept of wages and salaries includes the total wage bill by the establishments as reported. In addition, output is measured as value-added. Consequently the treatment of intermediate input is concealed. There is an implicit assumption that the share of intermediate input in gross output is nearly constant, which can be interpretated as that the elasticity of substitution between intermediate input and value-added is zero. Gross sales is excluded because detailed experimentation with Philippine manufacturing data by Sicat (1970) led to the conclusion that it matters little whether gross sales or value-added is used to measure output.Furthermore,value-added is the normal measure for output [Ferguson, 1965; Sicat 1970; Behrman, 1972; Jae Won Kim, 1984].

Estimation of the CES production function, elasticity of substitution and technical change under perfect competition

The estimation of the elasticity of substitution implies an analysis of the production function and its translation into an estimating form as shown by equation 3:19. The estimating form of this concept correlates labor productivity and the real wage rate. The former is explained by a number of variables such as technical progress, scale of output and changes in the wage rate. For example, the expansion of output would tend to lead to a furthur division of labor and to a higher labor productivity because of internal and external economies which depend on the structure of final demand, the "state of the art", supply of labor and other resources and the existing organisation of the industry.

A series of regression equations were fitted to time-series data 1963 - 1984 for fifty 5-digit industry groups in the Malaysian manufacturing sector. The first series of equations (Models A1 - A12) were applied to equation [3:19], to determine the elasticity of substitution between capital and labor based on competitive conditions in commodity and factor markets.

[MODEL A1] $\ln(\frac{VA}{L}) = a + b_1 \ln W_t + b_2 T + \epsilon_t$

[MODEL A2] $\ln(\frac{VA}{L}) = a + b_1 \ln W_t + b_2 \ln T + \epsilon_t$

[MODEL A3] $\ln(\frac{VA}{L}) = a + b_1 \ln W_t + b_2 (IPI) + \epsilon_t$

[MODEL A4) $\ln(\frac{RVA}{L}) = a + b_1 \ln (RW)_t + \epsilon_t$

[MODEL A5] $\ln(\frac{VA}{L}) = a + b_1 \ln(\frac{W}{L}) + b_2 \ln T + \epsilon_t$

[MODEL A6] $\ln(\frac{RVA}{L}) = a\ b_1 \ln(\frac{RW}{L}) + b_t \ln T + \epsilon_t$

[MODEL A7] $\ln(\frac{VA}{L}) = a + b_1 \ln(\frac{W}{L}) + b_2 \ln(\frac{VA}{L}) + b_3 \ln T + \epsilon_t$

[MODEL A8] $\ln(\frac{RVA}{L}) = a + b_1 \ln W_t + b_2 \ln(\frac{RVA}{L})_t + \epsilon_t$

[MODEL A9] $\ln(\frac{IVA}{L}) = a + b_1 \ln(\frac{IW}{L}) + b_2 \ln T + \epsilon_t$

[MODEL A10] $\ln(\frac{VA}{L}) = a + b_1 \ln(\frac{W}{L}) + b_2 \ln(\frac{VA}{L}) + b_3 \ln(W) + b_4 \ln T + \epsilon_t$

[MODEL A11] $\ln(\frac{VA}{L}) = a + b_1 \ln(\frac{W}{L}) + b_2 \ln(\frac{W}{L})^2 + b_3 \ln T + b_4 \ln T^2 + \epsilon_t$

[MODEL A12] $\ln(\frac{VA}{L}) = a + b_1 \ln W_t + b_2 \ln W_t^2 + b_3 \ln T + \epsilon_t$

where

VA	=	value added in current value
L	=	number of labor
W	=	wages and salaries in current value
T	=	time
RVA	=	value-added deflated with Consumer Price Index (CPI)
RW	=	wages and salaries deflated with CPI
IVA	=	value-added deflated with Industrial Production Index (IPI)
IW	=	wages and salaries deflated with IPI
ϵ_t	=	error term

Using the Time-Series Processing Package (TSP), models A1, A2, and A3 were estimated by Ordinary Least Squares (OLS) for fifty 5-digit Industry groups in the Malaysian manufacturing sector. The initial results show a number of problems including serial correlation, multicollinearity, and negative signs for the elasticity estimates in a large number of industry groups. Multicollinearity in Model A2 is however not serious. For most industry groups the elasticity estimate is significant and the covariance matrix between the dependent and independent variables is less than 0.5.

Models A1 and A3 are excluded because the coefficient of determination, R^2 is as low as 0.003 for a number of industries. Furthermore, the elasticity estimate is negative in forty-one out of fifty industry groups.

The next procedure is to test for the aptness of the OLS to be applied to the Malaysian data. A rigorous residuals test is carried out to determine linearity of the regression functions, constancy of variance, and normality of error terms as well as the presence of outliers, based on Model A2. These tests are carried out to ensure that the assumptions of the OLS are not violated. Scatter diagrams and plots of overall residuals against the explanatory variables and the dependent variables, the logarithm of value-added per worker and the logarithm of time are examined. Although there are some outliers outside two standard deviations from the mean, it was concluded that OLS will give efficient estimates of the elasticity of substitution between capital and labor for the 5-digit industry groups in the Malaysian manufacturing sector.

The next procedure was to fit OLS to models A4, A5, A6, A7, A8, A9, A10, A11 and A12. Model A5 is an adjustment of model A2, whereby the logarithm of value-added per labor is regressed againt the logarithm of wages and salaries per labor and the logarithm of time. Models A4 and A6 are replications of models A2 and A5 using data which is deflated by the consumer price index (CPI). Similarly, model A9 is a replication of model A5 with data which is deflated by the Industrial Production Index (IPI). Attempts to check for nonlinearity effects were made by fitting quadratic models A11 and A12. The Partial Adjustment Model A7 and Serial Correlation Model A10 were fitted to see whether there will be improvements in the results of the previous regression models. Finally a search procedure was carried out to determine the best fitting models.

Discussion of Empirical Results

There are a number of conceptual and data problems connected with the estimation of the elasticity of substitution using the CES production function. First, the classical procedure of estimation assumes an exogeneous determination of the wage rate. If the wage rate is not exogenously determined, the CES procedure yeilds a biased and inconsistent estimate of the elasticity even if returns to scale are constant and the wage rate is equal to the marginal product of labor. The exogenous determination of the wage rate is more likely to be a more relevant assumption for an international sample than for an aggregate time series within a single country [Feldstein, 1967]. Secondly, the data are assumed to represent points on the production frontier, that is, all firms are assumed to have adjusted fully to the prevailing factor prices within the observation period. Thirdly, no attempt is made to allow for variations in capacity utilization, or for fluctuations in the level of economic activity. In time series analysis, much of the variation in value-added may be attributed to different rates of capacity utilisation over a business cycle. Furthermore, during the period 1963 - 1984, a mild recession occurred in 1972 and a stronger one since 1980. This would impart a downward bias to the estimate of the elasticity of substitution, attributed to changes in the quality of labor services. Theoretically, in recession years, an increase in unemployment is normally accompanied by an increase in the quality of labor services since the more efficient workers will be retained. Thus value-added per worker tends to increase during periods of recession. Hence, our estimate of the time-series elasticity of substitution may be biased to an unknown extent.

Leaving aside the data and estimation problems, which are always present to a greater or lesser degree in any empirical work, our evaluation of the results hinge on how we view the causality between wages and capital - labor ratios. If there is evidence to show that capital - labor ratios are efficiently flexible and that entreprenuers do respond to factor price incentives, then the results of the regressions provide support for the view that making capital cheaper relative to labor tends to cause factor substitution toward greater capital-intensity. In Malaysian manufacturing sector, an average of 67.5% of local and foreign entreprenuers applied for tax incentives especially pioneer status between 1974 - 1984 [MIDA, Annual Reports]. This gives some credence to the view that Malaysian entreprenuers do respond to factor price incentives.

Models A1 - A12 are fitted to time series data covering 35 5-digit Malaysian manufacturing industries over the period 1963 - 1984, 5 industries over the period 1968 - 1984, and 10 industries over the period 1970 - 1984. The statistical search procedure shows Models A2, A5 and A6 as best fitting models to estimate the elasticity of substitution in Malaysian manufacturing industries. The results of models A2, A5 and A6 are given in Table 3.1.

Model A6 is finally chosen as the best regression equation based on both statistical and conceptual reasons. All three models are subjected to autocorrelative disturbances and are corrected using AR(1) method [Cochrane & Orcut 1949].

The AR(1) method provides efficient estimates of an equation whose disturbances display first order serial correlation, that is $u(t) = e(t) + rho^* u(t-1)$ [Rao & Griliches, 1969]. This method estimates rho from ordinary least squares residuals, transforms the dependent and independent variables so that the residuals from the transformed equation will be roughly serially uncorrelated, and then runs a regression using the transformed variables. This process is repeated until rho converges or maximum number of iterations is reached. The Cochrane - Orcutt procedure is asymptotically equivalent to the maximum likelihood procedure [Beach & MacKinnen, 1978].

In model A2, the regression estimated is

$$\ln\left(\frac{VA}{L}\right) = a + b_1 \ln W_t + b_2 \ln T + \epsilon_t$$

Table 3.1. Statistical performance of alternative models of the CES production function in Malaysian manufacturing sector, 1963-1984

INDUSTRY/MODEL	[MODEL A2]	[MODEL A5]	[MODEL A6]
Slaughtering Preparing & Perserving Meat	−.45[a] (−3.15) (6.14)	R^2 = .98 F = 152.1	−.26 (−.49) (.53)	R^2 = .86 F = 20.9	−.11 (−.22) (.50)	R^2 = .71 F = 8.3
Ice Cream Manufacturing	−0.04 (−.16) (.28)	R^2 = .94 F = 96.7	.65 (2.96) (.22)	R^2 = .92 F = 68.7	.65* (1.59) (.41)	R^2 = .74 F = 15.8
Manufacture of Other Dairy Products	.87* (6.22) (.14)	R^2 = .96 F = 134.2	.75* (1.47) (.51)	R^2 = .51 F = 3.42	.81 (1.56) (.52)	R^2 = .29 F = 1.39
Pineapple Canning	1.59 (2.13) (.78)	R^2 = .71 F = 14.1	1.10* (3.24) (.34)	R^2 = .75 F = 16.9	1.09* (2.52) (.43)	R^2 = .65 F = 10.3
Other Canning and Preserving of Fruits and Vegetable	−.34 (−.09) (.34)	R^2 = .57 F = 4.4	.75* (1.47) (.51)	R^2 = .51 F = 3.42	.81* (1.56) (.52)	R^2 = .29 F = 1.39
Coconut Oil Manufacturing	−.54 (−2.26) (.24)	R^2 = .58 F = 7.83	.83 1.59 .52	R^2 = .29 F = 2.36	1.04* 1.96 .52	R^2 = .65 F = 10.46
Palm Oil Manufacturing	.03 (.09) (.33)	R^2 = .73 F = 6.21	.59 .35 1.69	R^2 = .10 F = .27	1.36 .67 2.01	R^2 = .59 F = 3.37
Palm Kernel Oil Manufacturing	.24 (.52) (.46)	R^2 = .92 F = 27.3	.24 .24 .99	R^2 = .81 F = 9.91	.27 .27 1.007	R^2 = .62 F = 3.81
Vegetable and Animal Oils & Fats	.56* (4.05) (.14)	R^2 = .96 F = 146.3	1.32* 4.56 .29	R^2 = .91 F = 59.6	1.43* (4.15) (.35)	R^2 = .85 F = 32.6
Rice Milling	−.16 (−.85) (.18)	R^2 = .77 F = 18.5	.13 .23 .52	R^2 = .63 F = 9.8	.83 1.09 .76	R^2 = .58 F = 2.29
Biscuit Factories	.006 (.04) (.14)	R^2 = .98 F = 257.5	1.08* 10.22 .11	R^2 = .94 F = 91.69	.99* 3.38 .29	R^2 = .52 F = 5.96
Sugar Factories and Refineries	.38 (1.53) (.25)	R^2 = .35 F = 2.2	1.41* 5.38 .26	R^2 = .72 F = 10.12	1.45* 5.08 .28	R^2 = .68 F = 8.47
Manufacture of Cocoa, Chocolate and Sugar Confectionery	−.12 −.43 (.28)	R^2 = .96 F = 90.5	1.65* 2.33	R^2 = .78 F = 11.7	1.74* 2.77	R^2 = .56 F = 3.39
Ice Factories	.23 (1.69) (.14)	R^2 = .28 F = 2.20	.65* 2.96 .22	R^2 = .92 F = 68.67	.65* 1.59 .41	R^2 = .74 F = 15.8

Table 3.1. Continued

INDUSTRY/MODEL	[MODEL A2]	[MODEL A5]	[MODEL A6]
Coffee Factories	−.32* $R^2 =$ .98 (−1.82) $F = 229.3$ (.18)	.80* $R^2 =$ .95 2.71 $F = 64.59$ 1.30	.86* $R^2 =$ .71 .28 $F = 8.28$ .28
Meehoon, Noodles & Related Products	−.11 $R^2 =$ .99 −1.52 $F = 392.5$ .07	.34* $R^2 =$ .90 1.74 $F = 29.6$ .19	.33 $R^2 =$ .65 1.49 $F = 6.16$ .22
Manufacture of Prepared Animal Feeds	.69* $R^2 =$ .95 (4.71) $F = 120.2$ (.15)	.99* $R^2 =$ .95 14.12 $F = 112.8$ .07	.93* $R^2 =$ .72 7.05 $F = 14.5$ .13
Soft Drinks and Carbonated Beverages	.41* $R^2 =$ .97 (3.51) $F = 182.9$ (.12)	.75* $R^2 =$ .97 6.19 $F = 181.86$ .12	.65* $R^2 =$ .87 3.32 $F = 38.42$ .19
Tobacco Manufacturing	.08 $R^2 =$ .98 (.67) $F = 297.3$ (.12)	1.05* $R^2 =$ .92 6.10 $F =$ .17	1.04* $R^2 =$ .81 5.71 $F = 24.5$ .18
Manufacture of Leather & Leather Products	−.54 $R^2 =$ .94 6.26 $F = 11.23$ .11	.69* $R^2 =$ .77 6.56 $F = 18.49$ .11	.69* $R^2 =$ .85
Sawmilling	−.03 $R^2 =$ .98 (−.33) $F = 339.8$ (.09)	.59* $R^2 =$ .91 1.82 $F = 59.8$ .32	2.22* $R^2 =$ .58 3.05 $F = 7.73$ .73
Planning Mills & Joinery Works	.40* $R^2 =$ .98 (7.09) $F = 251.5$ (.06)	.52 $R^2 =$ .94 1.46 $F = 94.9$ .36	.44 $R^2 =$ .73 1.03 $F = 14.92$ .42
Manufacture of Furniture & Fixtures	−.09 $R^2 =$ .97 −(1.10) $F = 34.3$ (.26)	1.02* $R^2 =$ .89 8.23 $F = 28.02$ .12	1.02* $R^2 =$ .93 8.34 $F = 42.2$ .12
Clothing Factories	−.35 $R^2 =$ .98 (−1.64) $F = 178.9$ (.21)	−.24 $R^2 =$ .49 −.174 $F = 2.94$ .38	−1.27 $R^2 =$ .18 −.87 $F = .7$ 1.45
Manufacture of Paper & Paper Products	−.74 $R^2 =$ .96 (−1.06) $F = 76.3$ (.69)	1.38* $R^2 =$ .57 3.81 $F = 4.35$ .36	1.38* $R^2 =$ .35 3.83 $F = 1.8$ .36
Printing Publishing & Allied Industries	.61* $R^2 =$ .91 (3.37) $F = 60.0$ (.18)	−.27 $R^2 =$ .77 −.85 $F = 18.74$ .32	.88* $R^2 =$ .66 2.28 $F = 10.9$ .38
Manufacture of Basic Industrial Chemicals	−.28 $R^2 =$ .45 (−1.25) $F = 4.6$ (.23)	.83* $R^2 =$ .95 5.51 $F = 98.4$ .15	.82* $R^2 =$ .08 5.23 $F = 43.45$ .16
Manufacture of Chemical Fertiliser & Pesticides	−.04 $R^2 =$ .97 (−.74) $F = 214.4$ .05	.09* $R^2 =$ .64 .22 $F = 9.97$ .44	.31 $R^2 =$ .15 .57 $F = 1.07$ .55

Table 3.1. Continued

INDUSTRY/MODEL	[MODEL A2]	[MODEL A5]	[MODEL A6]
Manufacture of Paints, Varnishes and Lacquers	.21* $R^2 =$ .98 (1.38) F = 316.8 (.15)	1.06* $R^2 =$ .98 14.08 F = 243.4 .07	.57* $R^2 =$ .83 2.83 F = 28.4 .20
Manufacture of Drugs, Medicine & Pharmaceuticals	−.72* $R^2 =$ .96 (1.02) F = 137.1 (.71)	1.15* $R^2 =$ .70 3.24 F = 13.48 .35	.063 $R^2 =$ .55 .165 F = 6.98 .383
Manufacture of Soap & Cleaning Preparation	.302* $R^2 =$ .96 (1.81) F = 161.6 (.17)	.82* $R^2 =$ .89 3.56 F = 49.9 .23	1.10* $R^2 =$ .64 3.25 F = 10.2 .34
Manufacture of Perfumes, Cosmetics & Toileteries	.89* $R^2 =$ .75 (6.13) F = 17.7 (.15)	.64* $R^2 =$ .61 2.96 F = 8.78 .22	.59* $R^2 =$ .64 2.57 F = 10.25 .23
Petroleum Refineries	−1.31 $R^2 =$ .97 (−1.75) F = 120.3 (.75)	3.03 $R^2 =$ .50 1.32 F = 3.97 2.29	3.51* $R^2 =$ .30 (1.43) F = 1.74 2.45
Petroleum & Coal Products	.43* $R^2 =$ .96 (10.1) F = 72.3 (.04)	1.15* $R^2 =$ .77 3.88 F = 11.4 .29	1.12* $R^2 =$ .83 3.89 16.17 .29
Rubber Products	.96* $R^2 =$ .90 (20.7) F = 80.1 (.04)	.83* $R^2 =$ .93 (5.59) F = 67.9 (.15)	.88* $R^2 =$ .80 (5.45) F = 21.45 (.16)
Plastic Products	−.29 $R^2 =$ .96 (−.67) F = 136.8 (.15)	−.43* $R^2 =$ .94 −.12 F = 46.7 .37	−.15 $R^2 =$ .61 −.37 F = 4.68 .42
Pottery China & Earthernware	−.13 $R^2 =$ .95 (−1.05) F = 119.6 (.13)	1.58* $R^2 =$ .66 2.93 F = 10.79 .54	1.73* $R^2 =$ .35 1.46 F = 3.10 .19
Hydraulic Cement	.79* $R^2 =$ .91 (2.78) F = 59.5 (.29)	1.23* $R^2 =$ .83 2.26 F = 27.43 .55	2.72* $R^2 =$ .64 3.51 F = 10.16 .62
Cement & Concrete	−.08 $R^2 =$ .97 (−.19) F = 160.1	.32* $R^2 =$ .60 .54 F =	−.08 $R^2 =$ .22 −.12 F = 1.6
Iron Foundaries	−.10 $R^2 =$ .93 (−1.09) F = 67.1 (.26)	−.29 $R^2 =$ .82 1.25 F = 18.76 .23	−.61* $R^2 =$ .22 2.198 F = 1.49 .28
Non–Ferrous Metal Product	.29* $R^2 =$ .94 (4.63) F = 91.4 (.06)	−.39 $R^2 =$ .59 −1.75 F = 8.19 .23	−.27 $R^2 =$ .07 −.99 F = .48 .27
Wire Products Manufacturing	−.27 $R^2 =$ .92 .5 F .26	.31* $R^2 =$ .86 1.22 F = 35.46 .38	−.36 $R^2 =$ .70 −.92 F = 13.1

Table 3.1. Continued

INDUSTRY/MODEL	[MODEL A2]	[MODEL A5]	[MODEL A6]
Brass, Copper, Pewter & Alluminium Product	.50* $R^2 =$ .93 (3.29) F = 77.9 (.15)	.94* $R^2 =$ .94 8.96 F = 90.93 .11	.87* $R^2 =$ .73 4.33 F = 15.54 .19
Industrial Machinery & Parts	–.28 $R^2 =$ .45 (–1.25) F = 4.5 (.23)	.69* $R^2 =$ .70 6.66 F = 13.49 .10	.70* $R^2 =$ .86 7.16 F = 34.4 .09
Electrical Machinery, Apparatus & Appliances	–.34 $R^2 =$ .85 (–.50) F = 50.6 (.21)	.49* $R^2 =$ .65 2.96 F = 6.09 .17	.48* $R^2 =$ .40 2.76 F = 2.19 .18
Shipbuilding, Boatmaking & Repairing	.05 $R^2 =$ .98 (–.50) F = 367.8 (.12)	.86* $R^2 =$ .90 3.68 F = 51.02 .23	.75* $R^2 =$ .80 3.31 F = 22.4 .23
Manufacturing of Motor Vehicle Bodies	.59* $R^2 =$ .98 (12.79) F = 292.5	.53* $R^2 =$ .98 2.52 F = 274.7	.50* $R^2 =$ .92 2.37 F = 65.68
Manufacture of Motor Vehicle Parts & Accessories	.18 $R^2 =$ .98 (1.36) F = 306.8 (.13)	1.30* $R^2 =$ .95 8.75 F = 99.9 .15	–.41 $R^2 =$ .87 –.95 F = 37.65 .43
Manufacture & Assembly of of Bicycle	.08 $R^2 =$ .08 (.28) F = .54 (.28)	.89* $R^2 =$ .59 2.90 F = 8.28 .31	1.05* $R^2 =$ .39 3.23 F = 3.6 .33
Manufacture of Professional & Scientific Equipment	.69 $R^2 =$ .22 (1.00) F = .9 (.69)	–.17* $R^2 =$ .24 –1.23 F = 1.04 .14	–.17 $R^2 =$ .48 –1.18 F = 3.12 .14

[a]For each industry group, the first row shows the estimate of the elasticity of substitution (σ) the second row shows the t-statistics for the elasticity estimate the third row shows the standard error for the elasticity estimate
*implies significant at 1% level of significance.

Except for seven industries, the coefficient of determination R^2 is very high and highly significant. The logarithms of wages and salaries explain the greater amount of total variation in value-added in fifty-two industries. There is however serious multicollinearity in thirty-eight industries, including tobacco manufacturing, sawmilling, cement and concrete manufacturing, and others. As seen in Table 3.1, the R^2 ranges between 0.28 - 0.98. However, the t-ratios for majority of industries, as shown in table 3.1 are very low, and none of the partial regression coefficients are individually statistically significant. It is suspected that with time series data 1963 - 1984, both value-added per employee and wages and salaries are moving in the same direction. One way of optimising this dependence and correcting for multicollinearity is by transforming the explanatory variables. The logarithm of value-added per man is regressed against the logarithm of wages and salaries per man and the logarithm of time as shown in Model A5.

$$[\text{A5}] \quad \ln\left(\frac{VA}{L}\right) = a + b_1 \ln\left(\frac{W}{L}\right) + b^2 \ln T + \epsilon\, t$$

Model A5 shows a great improvement in the results. The coefficient of determination R^2 is still statistically significant. However, it is not as high as in Model A2. Based on conventional t-statistics, the partial regression coefficients are statistically significant at 5% significance level in 42 industries.

M Nerlove (1967) and others have shown that when wages are not highly negatively correlated with prices, failure to deflate data on output and wages will tend to bias upwards the estimated elasticity of substitution. Furthermore, if time series data are undeflated from an inflationary situation, variations in the rate of inflation will result in further bias of the estimates of the elasticity of substitution upwards. In an attempt to reduce such potential sources of bias, Model A5 was re-estimated, by Models A6 and A9 respectively. Model A6 is

$$\ln\left(\frac{VAR}{L}\right) = a + b_1 \ln\left(\frac{WR}{L}\right) + b_2 \ln T + \epsilon_t$$

where $\left(\dfrac{VAR}{L}\right)$ is value added per labor deflated by the Consumer Price Index (CPI)

$\left(\dfrac{WR}{L}\right)$ is wages and salaries per labor deflated by the CPI

T is time

and Model A9 is

$$\ln\left(\dfrac{VAI}{L}\right) = a + b_1 \ln(WI) + b_2 \ln T + \epsilon_t$$

where $\left(\dfrac{VAI}{L}\right)$ is value added per labor deflated by the Industrial Production Index (IPI)

$\left(\dfrac{WI}{L}\right)$ is wages and salaries per labor deflated by the (IPI)

T is time

Theoretically, deflating the time series data will yield efficient and unbiased estimators. A further advantage of using deflated time series is that extreme observations will have less effect on the estimation [Kuh & Meyer, 1955] and will reduce the bias due to those outlying observations.

Statistically, we can observe the impact of deflating the series by comparing the results of models A5 and A6. As shown in Table 3.1, the elasticity of substitution between capital and labor is reduced in 31 cases basing on models A5 and A6. This can be interpreted as a reduction in the various biases present in Model A5 which are due to the data as well as the postulated model itself [Draper & Smith 1981]. Statistically, model A5 exhibits superior results in terms of R^2, adjusted R^2, t-statistics and minimum standard errors of the elasticity coefficient. The elasticity coefficient is statistically significant at the 5% significance level in 42 industry groups in model A5. However, it is statistically significant at a similar level in only 32 industry groups in Model A6. We are thus faced with the dilemma that model [5] with superior statistical results is biased while model A6 with not as good statistical results contains lesser bias. Faced with choosing between model A5 which yields good statistical results but with lesser meaning and model A6 which yields more meaningful though not as good statistical results, it seems clear that the latter alternative is the better choice.

Model A9 is rejected because the industrial production index is available only from 1970, thus reducing the time series to 15 observations. Model A6 is therefore chosen as the most appropriate estimating form to analyse substitution possibilities, technical change, structural adjustments and technological change in the Malaysian manufacturing industries, based on the CES Production functions.

The estimates of the elasticity of substitution between capital and labor in Malaysian manufacturing sector based on model [A6] are not free from bias. The simultaneity problem, as shown by Madalla and Kadane (1968), biases downward the elasticity estimates. Secondly, errors of measurement in the variables as discussed earlier result in asymptotically downward biases in estimates of σ from Ordinary Least Squares (OLS) procedures. Thirdly, omitting inter-industry differentials in the quality of labor will bias the estimate of the elasticity towards unity [Griliches, 1967]. The choice of Model [A6] is that it contains the least bias compared to other estimates.

Elasticity of substitution and technical change under perfect competition, 1963 - 1984

Based on Model A6, the time-series estimates of the elasticity of substitution are positive in forty one industries indicating that efficient factor substitutability is possible. The coefficient denoting the elasticities are statistically significant at 5% for thirty-five industries. More than half of the coefficients (34 out 50) are numerically less than unity, while in sixteen industries, the elasticity of substitution exceed unity. In 48.0 percent of the cases, the value of elasticity is less than 0.80. In 20.0 percent of the cases it is greater than 1.10, while for the remaining 32.0 percent of the cases, it lies between 0.80 and 1.10. The time-series estimates of the elasticity of substitution of 5- digit Industrial Classification of Malaysian manufacturing industries ranged from 3.51 for petroleum refining to - 1.27 for clothing manufacturing. Table 3.3 shows the ranking of industries according to the elasticity of substitution between capital and labor.

The numerical value of elasticity of substitution is unity in the Cobb-Douglas case and zero in the fixed-proportions case, therefore a t-test is needed of whether or not σ is significantly different from unity. The results of the hypothesis $\sigma = 1$ and $\sigma = 0$ are presented in Table 3.4.

Twenty-six estimates which show positive elasticity of substitution are significantly different from zero at 95% level of confidence. This evidence should discredit the notion of fix propotions in industrial sector

Table 3.2. Time series estimates of the elasticity of substitution of Malaysian manufacturing industries, 1963-1984

	Elasticity of Substitution	Standard Error	R^2
Slaughtering, Preparing and Preserving Meat	−0.11	−0.50	0.71
Ice cream Manufacturing	0.65*	0.41	0.74
Manufacture of Other Dairy Products	0.41	0.52	0.29
Pineapple Canning	1.09*	0.43	0.65
Other Canning and Preserving of Fruits and Vegetable	0.81*	0.52	0.29
Coconut Oil Manufacturing	1.04*	0.52	0.65
Palm Oil Manufacturing	1.36	2.01	0.59
Palm Kernel Oil Manufacturing	0.27	1.01	0.62
Vegetable and Animal Oils Fats	1.43*	0.35	0.85
Rice Milling	0.83	0.76	0.58
Biscuit Factories	0.99*	0.29	0.52
Sugar Factories & Refineries	1.45*	0.28	0.68
Manufacture of Cocoa, Chocolate and Confectionery	1.74*	0.63	0.56
Ice Factories	1.65*	0.41	0.74
Coffee Factories	0.86*	0.28	0.71
Meehoon & Noodles and Related Products	0.33	0.22	0.65
Manufacture of Prepared Animal Feeds	0.93*	0.13	0.72
Soft Drinks and Carbonated Beverages	0.65*	0.19	0.87
Tobacco Manufacturing	1.04*	0.18	0.81
Manufacture of Leather and Leather Products	0.69*	0.11	0.85
Sawmilling	2.22*	0.73	0.58
Planning Mills and Joinery Works	0.44*	0.42	0.73
Manufacture of Furniture and Fixtures	1.02*	0.12	0.93
Clothing Manufacturing	−1.27	1.45	0.18
Manufacture of Paper and Paper Products	1.38*	0.36	0.35
Printing, Publishing and Allied Industries	0.88*	0.38	0.66
Manufacture of Basic Industrial Chemicals	0.82*	0.16	0.88
Manufacture of Chemical Fertiliser and Pesticides	0.31	0.55	0.15
Manufacture of Chemical Fertiliser and Lacquers	0.57*	0.20	0.83
Manufacture of Drugs Medicine and Pharmacueticals	0.06	0.38	0.55
Manufacture of Soaps of Cleaning Preparation	1.10*	0.34	0.64
Manufacture of Perfumes, Cosmetics and Toiletteries	0.59*	0.23	0.64

Table 3.2. Continued

	Elasticity of Substitution	Standard Error	R^2
Petroleum Refineries	3.51*	2.45	0.30
Petroleum and Coal Products	1.12*	0.29	0.83
Rubber Products	0.88*	0.16	0.80
Plastic Products	−0.15	0.42	0.61
Pottery China and Earthernware	1.73*	0.19	0.35
HydraulicCement	2.72*	0.62	0.64
Cement and Concrete	−0.08	0.66	0.22
Primary Iron & Steel Industries	−0.61*	0.28	0.22
Non Ferros Metal Products	−0.27	0.27	0.07
Wire Products Manufacturing	−0.36	0.38	0.70
Brass, Copper and Alluminium Products	0.87*	0.19	0.73
Industrial Machinery and Parts	0.70*	0.09	0.86
Electrical Machinery, Apparatus and Appliances	0.48*	0.18	0.40
Shipbuilding, Boatmaking and Repairing	0.75*	0.23	0.80
Manufacture of Motor Vehicle Bodies	0.50*	0.21	0.92
Manufacture of Motor Vehicles Parts and Accessories	−0.41	0.43	0.87
Manufacture and Assembly of Bicycles	1.05*	0.33	0.39
Manufacture of Professional and Scientific Equipment	−0.17	0.14	0.48

*Significant at 5% level.

Table 3.3. Ranking of elasticity of substitution in Malaysian manufacturing industries, 1963-1984

	Elasticity of Substitution (σ)
Petroleum Refineries	3.51*
Hydraulic Cement	2.72*
Sawmilling	2.22*
Cocoa, Chocolate and Confectionery	1.74*
Manufacture of Pottery, China and Earthernware	1.73*
Sugar Factories & Refineries	1.45*
Vegetable and Animal Oils Fats	1.43*
Paper and Paper Products	1.38*
Palm Oil Manufacturing	1.36
Petroleum and Coal Products	1.12*
Soap and Detergents	1.10*
Pineapple Canning	1.09*
Ice Factories	1.06*
Bicycle Manufacturing	1.05*
Tobacco Manufacturing	1.04*
Coconut Oil Manufacturing	1.04*
Furniture and Fixtures	1.02*
Biscuit Factories	0.99*
Prepared Animal Feeds	0.93*
Rubber Products manufacturing	0.88*
Brass, Copper and Alluminium Products	0.87*
Coffee Factories	0.86*
Rice Milling	0.83
Industrial Chemical	0.82*
Other Canning	0.81*
Shipbuilding, Boatmaking and Repairing	0.75*
Industrial Machinery and Parts	0.70*
Leather and Leather Products	0.69*
Soft Drinks and Carbonated Beverages	0.65*
Ice cream Manufacturing	0.65*
Perfumes, Cosmetics and Toiletteries	0.59*
Pains, Varnishes and Lacquer Industries	0.57*
Motor Vehicle Bodies	0.50*
Electrical Machinery and Apparatus	0.48*
Planning Mills and Joinery Works	0.44*
Dairy Products	0.41
Meehoon and Noodles	0.33

Table 3.3. Continued

	Elasticity of Substitution (σ)
Chemical Fertiliser	0.31
Palm Kernel Oil Manufacturing	0.27
Medicinal and Pharmacueticals Products	0.06
Cement and Concrete	− 0.08
Slaughtering, Preparing and Preserving Meat	− 0.11
Plastic Products	− 0.15
Manufacture of Professional and Scientific Equipment	− 0.17
Non Ferros Metal Products	− 0.27
Wire Products Manufacturing	− 0.36
Motorcar Part and Accessories	− 0.41
Primary Iron & Steel Industries	− 0.61*
Clothing Manufacturing	− 1.27

* Implies Significant at 5% significance level.

of developing countries. The belief that capital- intensive manufacturing processes similar to those found in developed countries (current western technology) are the only alternatives for developing countries such as Malaysia was quite strong in the 1950's and the 1960's. The major argument in favor of them was that they were simply more efficient than more labor-intensive alternatives. The latter, it was claimed, would always use more labor and more capital per unit of output than would the process with the high capital-labor ratio. Thus, although alternatives might exist in a technical sense, they would always be found to be inferior [Samir Amin, 1969;P.D Ady 1971: L.Barber, 1969].

The evidence presented in Table 3.4 shows that at least twenty-six industries in Malaysian manufacturing sector with positive elasticities are not characterised by fixed proportions. In eight industries (sugar factories and refineries, soft drinks and carbonated beverages, manufacture of leather and leather products, sawmilling, paints, varnishes and lacquer industries, perfumes and cosmetics industries, hydraulic cement industries and industrial machinery and parts), the elasticity of substitution is statistically greater than unity, indicating that the general form of the CES production function is the only suitable model. Given the time series data 1963 - 1984, these industries have shown a small degree of factor substitutability in the Malaysian manufacturing sector.

However, as shown in Table 3.4, more than half of the industries, constituting twenty-seven out of fifty (54.0%), have elasticities that are not statistically different from unity. In these cases, the CES function, with unitary elasticity of substitution between capital and labor is reduced to the Cobb-Douglas form, thus the latter model may provide a more suitable statistical representation of the underlying production function for these industries.

The impact of the substitution elasticity (σ) upon factor composition and the factor shares is well-known. In the process of growth with technical progress, the labor absorption depends crucially on the substitution elasticity. In the case of $\sigma = 1$, neither labor nor capital - augmenting technical progress affects the labor intensity, assuming a constant factor price ratio. However, when $\sigma > 1$, a labor - augmenting progress increases the labor intensity and consequently the demand for labor while a capital-augmenting progress reduces the labor absorption. The reverse holds for $\sigma < 1$. Thus a discussion of the relationship between elasticity of substitution and the rate of technical progress can provide a better understanding of the Malaysian manufacturing sector.

Table 3.5 shows the rate of technical progress (λ) corresponding to significant estimates of the elasticity of substitution (σ) in the Malaysian manufacturing industries. Between 1963 - 1984, only sugar

Table 3.4. Tests of significance of the elasticity of substitution of Malaysian manufacturing industries[a]

	Elasticity of Substitution (σ)	df	Significance Level of σ different from 1	Significance Level of σ different from 0
Slaughtering, Preparing and Preserving Meat	−0.11	12	5%	n.s
Ice cream Manufacturing	0.65	19	n.s	5%
Manufacture of Other Dairy Products	0.41	19	n.s	n.s
Pineapple Canning	1.09	19	n.s	5%
Other Canning and Preserving of Fruits and Vegetable	0.81	12	n.s	5%
Coconut Oil Manufacturing	1.04	19	n.s	5%
Palm Oil Manufacturing	1.36	12	n.s	5%
Palm Kernel Oil Manufacturing	0.27	12	5%	n.s
Vegetable and Animal Oils Fats	1.43	19	n.s	5%
Rice Milling	0.83	19	n.s	n.s
Biscuit Factories	0.99	13	n.s	5%
Sugar Factories & Refineries	1.45	13	n.s	5%
Manufacture of Cocoa, Chocolate and Confectionery	1.74	12	n.s	5%
Ice Factories	0.65	19	n.s	5%
Coffee Factories	0.86	12	n.s	5%
Meehoon & Noodles and Related Products	0.33	12	5%	n.s
Manufacture of Prepared Animal Feeds	0.93	19	n.s	5%
Soft Drinks and Carbonated Beverages	0.65	19	n.s	5%
Tobacco Manufacturing	1.04	19	n.s	5%
Manufacture of Leather and Leather Products	0.69	12	n.s	5%
Sawmilling	2.22	19	5%	5%
Planning Mills and Joinery Works	0.44	19	5%	5%
Manufacture of Furniture and Fixtures	1.02	12	n.s	5%
Clothing Factories	−1.27	12	5%	n.s
Manufacture of Paper and Paper Products	1.38	12	n.s	5%
Printing, Publishing and Allied Industries	0.88	19	n.s	5%
Manufacture of Basic Industrial Chemicals	0.82	19	n.s	5%
Manufacture of Chemical Fertiliser and Pesticides	0.31	19	5%	n.s
Manufacture of Chemical Fertiliser and Lacquers	0.57	19	5%	5%
Manufacture of Drugs Medicine and Pharmacueticals	0.06	19	5%	n.s
Manufacture of Soaps of Cleaning Preparation	1.10	19	5%	5%

Table 3.4. Continued

	Elasticity of Substitution (σ)	df	Significance Level of σ different from 1	0
Manufacture of Perfumes, Cosmetics and Toiletteries	0.59	12	5%	5%
Petroleum Refineries	3.5114	5%	5%	
Petroleum and Coal Products	1.12	12	n.s	5%
Rubber Products	0.88	19	n.s	5%
Plastic Products	−0.15	12	5%	n.s
Pottery China and Earthernware	1.73	19	n.s	5%
HydraulicCement	2.72	19	5%	5%
Cement and Concrete	−0.08	19	5%	n.s
Primary Iron & Steel Industries	−0.61	12	5%	5%
Non Ferros Metal Products	−0.27	19	5%	n.s
Wire Products Manufacturing	−0.36	19	5%	n.s
Brass, Copper and Alluminium Products	0.87	19	n.s	5%
Industrial Machinery and Parts	0.70	19	5%	5%
Electrical Machinery, Apparatus and Appliances	0.48	12	5%	5%
Shipbuilding, Boatmaking and Repairing	0.75	19	5%	5%
Manufacture of Motor Vehicle Bodies	0.50	19	5%	5%
Manufacture of Motor Vehicles Parts and Accessories	−0.41	19	5%	5%
Manufacture and Assembly of Bicycles	1.05	19	n.s	5%
Manufacture of Professional and Scientific Equipment	−0.17	12	5%	n.s

[a]All tests $\sigma = 0$ and $\sigma = 1$ are at 5% significance level.

Table 3.5. Time series estimates of the elasticity of substitution and the rate of technical progress in Malaysian manufacturing industries, 1963-1984

	Elasticity of Substitution (σ)	Rate of Technical Progress (λ)
Slaughtering, Preparing and Preserving Meat	−0.11	0.13
Ice cream Manufacturing	0.65*	0.89
Manufacture of Other Dairy Products	0.41	0.42
Pineapple Canning	1.09*	1.16
Other Canning and Preserving of Fruits and Vegetable	0.81*	0.66
Coconut Oil Manufacturing	1.04*	1.33
Palm Oil Manufacturing	1.36	0.39
Palm Kernel Oil Manufacturing	0.27	1.52
Vegetable and Animal Oils Fats	1.43*	1.79
Rice Milling	0.83	2.18
Biscuit Factories	0.99*	0.99
Sugar Factories & Refineries	1.45*	−1.11
Manufacture of Cocoa, Chocolate and Confectionery	1.74*	−1.08
Ice Factories	1.65*	1.05
Coffee Factories	0.86*	0.86
Meehoon & Noodles and Related Products	0.33	1.08
Manufacture of Prepared Animal Feeds	0.93*	0.98
Soft Drinks and Carbonated Beverages	0.65*	0.94
Tobacco Manufacturing	1.04*	1.00
Manufacture of Leather and Leather Products	0.69*	0.58
Sawmilling	2.22*	1.15
Planning Mills and Joinery Works	0.44*	0.67
Manufacture of Furniture and Fixtures	1.02*	0.85
Clothing Manufacturing	−1.27	0.38
Manufacture of Paper and Paper Products	1.38*	0.69
Printing, Publishing and Allied Industries	0.88*	1.53
Manufacture of Basic Industrial Chemicals	0.82*	0.97
Manufacture of Chemical Fertiliser and Pesticides	0.31	0.31
Manufacture of Chemical Fertiliser and Lacquers	0.57*	0.63
Manufacture of Drugs Medicine and Pharmacueticals	0.06	−0.21
Manufacture of Soaps of Cleaning Preparation	1.10*	1.16
Manufacture of Perfumes, Cosmetics and Toiletteries	0.59*	0.99
Petroleum Refineries	3.51*	4.22
Petroleum and Coal Products	1.12*	0.65
Rubber Products	0.88*	1.53

Table 3.5. Continued

	Elasticity of Substitution (σ)	Rate of Technical Progress (λ)
Plastic Products	−0.15	0.33
Pottery China and Earthernware	1.73*	0
HydraulicCement	2.72*	1.46
Cement and Concrete	−0.08	0.11
Primary Iron & Steel Industries	−0.61*	1.04
Non Ferros Metal Products	−0.27	−0.27
Wire Products Manufacturing	−0.36	0.92
Brass, Copper and Alluminium Products	0.87*	0.93
Industrial Machinery and Parts	0.70*	0.72
Electrical Machinery, Apparatus and Appliances	0.48*	0.46
Shipbuilding, Boatmaking and Repairing	0.75*	1.56
Manufacture of Motor Vehicle Bodies	0.50*	0.33
Manufacture of Motor Vehicles Parts and Accessories	−0.41	0.59
Manufacture and Assembly of Bicycles	1.05*	0.79
Manufacture of Professional and Scientific Equipment	−0.17	2.31

*Significant at 5% level.

refineries and factories experienced a negative rate of technical progress. The average annual rate of technical progress is 0.95 percent. Fifteen industries (hydraullic cement, sawmilling, cocoa, chocolate and confectionery, vegetable, oil and fats, soaps and detergents, pineapple canning, ice factories, architectural metal products, tobacco manufacturing, biscuit factories, prepared animal feeds, rubber products manufacturing, industrial chemicals and perfumes and cosmetics industries) experienced technical progress higher than the average rate. On the other hand, sugar refineries and factories show a negative rate of technical progress. The majority of industries however experience a relatively small rate of technical progress, well below average of 0.95.

It is noted that the industries which experience technical progress higher than the average rate are those with elasticities which are greater than unity. These industries such as hydraullic cement, sawmilling, cocoa, chocolate and confectionery are however capital-intensive industries (see Appendix). Thus, even with relatively higher but capital-augmenting technical progress, labor absorption is low.

Comparison of the CES estimates to other estimates of the elasticity of substitution.

Table 3.6 shows a comparison of the time series estimates of the elasticity of substitution between capital and labor in 5-digit Malaysian industries for the period 1963 - 1984 and cross-section estimates based on a 1974 survey of 338 manufacturing establishments in West Malaysia. Both cross-section estimates are derived from the CES production. The ACMS$^\sigma$ estimate is based on the equation,

$$\ln\left(\frac{VA}{L}\right) = a + \sigma \ln\left(\frac{W}{L}\right)$$

and the DIWAN$^\sigma$ estimate is based on the equation

$$\ln\left(\frac{C}{L}\right) = b + \sigma \left(\frac{W}{L}\right)$$

where C   is capital

L   is labor

TABLE 3.6. COMPARISON OF ALTERNATIVE ESTIMATES OF ELASTICITY OF SUBSTITUTION BASED ON CES PRODUCTION FUNCTION APPROACH

	1963–1984	1974	
	σCES[a]	σACMS[b]	σDIWAN[c]
Slaughtering, Preparing, and Preserving Meat	−0.11	−	−
Ice cream Manufacturing	0.65*	−	−
Manufacture of Other Dairy Products	0.41	1.04	0.89
Pineapple Canning	1.09*	−	−
Other Canning and Preserving of Fruits and Vegetable	0.81*	−	−
Coconut Oil Manufacturing	1.04*	−	−
Palm Oil Manufacturing	1.36	−	−
Palm Kernel Oil Manufacturing	0.27	−	−
Vegetable and Animal Oils Fats	1.43*	0.26	1.40
Rice Milling	0.83	0.43	0.73
Biscuit Factories	0.99*	0.51	0.54
Sugar Factories & Refineries	1.45*	−	−
Manufacture of Cocoa, Chocolate and Confectionery	1.74*	0.92	1.41
Ice Factories	1.65*	0.99	0.47
Coffee Factories	0.86*	0.91	−0.004
Meehoon & Noodles and Related Products	0.33	1.07	0.73
Manufacture of Prepared Animal Feeds	0.93*	0.50	0.14
Soft Drinks and Carbonated Beverages	0.65*	1.00	−
Tobacco Manufacturing	1.04*	1.23	1.78
Manufacture of Leather and Leather Products	0.69*	−	−
Sawmilling	2.22*	0.79	0.22
Planning Mills and Joinery Works	0.44*	0.62	0.23
Manufacture of Furniture and Fixtures	1.02*	−	−
Clothing Manufacturing	−1.27	0.83	0.29
Manufacture of Paper and Paper Products	1.38*	0.92	1.41
Printing, Publishing and Allied Industries	0.88*	0.76	0.08
Manufacture of Basic Industrial Chemicals	0.82*	−	−
Manufacture of Chemical Fertiliser and Pesticides	0.31	−	−
Manufacture of Chemical Fertiliser and Lacquers	0.57	0.64	0.82
Manufacture of Drugs Medicine andPharmacueticals	0.06	1.24	1.05
Manufacture of Soaps of Cleaning Preparation	1.10*	1.12	1.06
Manufacture of Perfumes, Cosmetics and Toiletteries	0.59*	1.31	1.09
Petroleum Refineries	3.51*	−	−
Petroleum and Coal Products	1.12*	−	−
Rubber Products	0.88*	1.06	0.41
Plastic Products	−0.15	0.84	0.74

TABLE 3.6. Continued

	1963–1984 σCES[a]	1974 σACMS[b]	σDIWAN[c]
Pottery China and Earthernware	1.73*	1.32	−0.19
HydraulicCement	2.72*	–	–
Cement and Concrete	−0.08	0.85	1.32
Primary Iron & Steel Industries	−0.61*	−0.04	−0.68
Non Ferros Metal Products	−0.27	–	–
Wire Products Manufacturing	−0.36	0.55	0.54
Brass, Copper and Alluminium Products	0.87*	0.99	0.97
Industrial Machinery and Parts	0.70*	0.51	0.09
Electrical Machinery, Apparatus and Appliances	0.48*	–	–
Shipbuilding, Boatmaking and Repairing	0.75*	0.23	1.36
Manufacture of Motor Vehicle Bodies	0.50*	–	–
Manufacture of Motor Vehicles Parts and Accessories	−0.41	0.64	0.55
Manufacture and Assembly of Bicycles	1.05*	0.93	1.33
Manufacture of Professional and Scientific Equipment	−0.17	–	–

[a] Elastisities derived from Model [A6] of this study
[b] Elasticities derived by Hoffman & Tan (1980) based on
Cross–sectional data 1974: $\text{Log}\left(\frac{VA}{L}\right) = a + b \log\left(\frac{W}{L}\right)$.
[c] Elasticities derived by Hoffman & Tan (1980) based on
cross–sectional data 1974: $\text{Log}\left(\frac{K}{L}\right) = a + b \log\left(\frac{W}{L}\right)$.
*Significant at 5% level.

W is wages and salaries

VA is value-added

σ is elasticity of substitution between capital and labor

It is striking to observe that there are some discrepancies between the values of the elasticity of substitution obtained for each industry, when this parameter is estimated with cross-sectional data or with time-series data. This is to be expected as pointed out by Nerlove (1967), "even slight variations in the period or concepts tend to produce drastically different estimates of the elasticity".

Comparing industry-wise the values of estimated elasticity of substitution (columns 1, 2, and 3 in Table 3.6), we note that in the majority of cases, the time-series value of this parameter came out invariably lower than their cross-sectional counterparts. We may observe that this pattern fits entirely with the general experience in this field.

Another interesting point to note is that the cross-sectional estimates using ACMS$^\sigma$ range between 1.62 and -0.03. While there are discrepancies in industry estimates, all the estimates seem to cluster around a similar range. Both cross-section and time-series estimates give the impression of generally low substitution possibilities between factors capital and labor in Malaysian manufacturing sector.

The estimates of the elasticity of substitution in Malaysian manufacturing sector gain further credence by comparing them with estimates in other developing countries. Ferguson's estimates for US manufacturing industries ranged from 0.24 to 1.30. Two sets of estimates of σ exists for the Peruvian manufacturing sector by industry. One set, obtained by Claque (1969) under alternative assumptions regarding capital inputs and interest rates, contains estimates which range between 0.125 and 1.106. Berhman (1972) presents estimates for Chilean manufacturing sector of 0.21 in the short run and 0.76 in the long run. Recent estimation of elasticities of substitution in Korean manufacturing industries [Jae Won Kim, 1984] shows a similar clustering between 0.01 and 0.89. Thus, except for Lianos's estimates for Greek manufacturing industries which range between -10.111 and 15.873, the majority of time-series estimates cluster between 0.1 and 1.3. Our estimates of the elasticity of substitution between capital and labor in Malaysian manufacturing industries cluster between 0.1 and 1.5. The average estimate is 0.72.

It is interesting to note the consistencies and similarities in the estimates of elasticities across countries. The time-series estimates of the elasticity of substitution in Greek manufacturing industries show negative elasticities for food manufacturing, electrical machinery appliances and transport equipment, and relatively low elasticities for textiles, footwear, leather products, rubber and plastic products manufacturing [Lianos, 1975]. Similarly, Ferguson (1965) found relatively lower elasticities in food, textiles, rubber and plastics, leather products, stone, clay and glass, and professional instruments manufacturing industries. Both Lianos and Ferguson reported the highest elasticities for petroleum manufacturing of 12.658 and 1.30 in Greek and American manufacturing industries respectively.

Thus, although the comparison of the estimates of the elasticity of substitution are drawn from different sample bases, different levels of aggregation and different estimating equations, there seems to be a consistent trend in the results of the time-series estimates. Strict comparability of estimates for policy purposes however would require strict comparability of both the treatment of data and the estimation procedures. Although no generalisation can be made about the size of the elasticity of substitution in the developing countries, the Malaysian estimates presented in this study are comparable to the time-series estimates in other countries based on the CES production functions.

Given the regression results, what can one conclude about the estimates of the elasticity of substitution in the Malaysian manufacturing sector?. The analysis based on the CES production function shows that the estimates cluster between 0.1 and 1.5. These estimates are however subjected to a number of biases in the specification of variables as well as in their estimation. Considering these biases, one tend to conclude that the actual elasticities are probably lower than the calculated elasticities.

An alternative method based on the duality between production and cost functions, can be used to estimate the elasticity of substitution. An attempt is made in the next chapter to determine the elasticity of substitution between capital and labor in the Malaysian manufacturing sector using the translog cost function.

CHAPTER IV.
CAPITAL-LABOR SUBSTITUTION IN MALAYSIAN MANUFACTURING USING TRANSLOG COST FUNCTION APPROACH

Methodology

From the neo-classical production function and the assumption of cost minimisation, factor demands are based on the necessary conditions for optimisation. These are readily obtained for simple production functions. However, factor demands are difficult to determine when the production technology is more complex. Furthermore, it is not possible to estimate the CES production function either in its original or logarithmic form because the bracketted term contains two of the parameters to be measured. It is preferable therefore, to study the factors of production or the factor demands in a manner which preserves the complexity of the structure of input decisions, yet simplifies the derivation. The duality between production and cost functions has provided this alternative. Varian (1978) and Diewert (1974) illustrate that the cost function contains all of the information about the production technology present in the conventional production function. Furthermore, given the regularity conditions, every cost function implies a well-behaved production technology. Thus, a logical approach in factor demand analysis is to proceed directly to a cost function without prior regard to a functional form for the production technology.

The theoretical and empirical model of the translog cost function.

Given a production function of the form

$$Y = f(K, L)$$

which summarises the underlying technology, where

    Y = gross output

    K = capital input

    L = labor input

Given the regularity conditions, and assuming factor prices (P_K) and (P_L) and output levels (Y) are exogenously determined, the theory of duality between production and cost functions states that by minimising cost, a cost function will be derived as follows:

$$l = P_KK + P_LL + [F(K,L) - Y]$$

$$\frac{\delta l}{\delta K} = P_K - F_K = 0$$

$$\frac{\delta l}{\delta L} = P_L - F_L = 0$$

$$\frac{\delta l}{\delta \theta} = F(K, L) - Y) = 0$$

Assuming second order conditions are not violated, the three first order conditions solve simultaneously the optimal factor demands

$$K^* (P_K, P_L, Y)$$

$$L^* (P_K, P_L, Y)$$

where K^* and L^* are optimal capital and labor inputs.

The Cost function is derived as

$$C^*(P_K, P_L, Y) = P_KK^*(P_K, P_L, Y) + P_LL^*(P_K, P_L, Y) \qquad (4:1)$$

Equation (4:1) is a general functional form. For purposes of empirical estimation, it is necessary to specify an explicit functional form for C^*. In this study, a popular cost function is estimated with Malaysian manufacturing data.

Christensen, Jorgensen and Lau (1973) propose the translog as an approximation to unknown cost or production functions by expressing them as a second-order polynomial in logarithms of input prices and output. Following Diewert (1974), the translog cost function which satisfies certain regularity conditions does correspond to a well-behaved production technology.

The translog cost function is expressed as

$$\ln C^*(P_K, P_L, Y) = a_o + a_y \ln Y + \theta/2 \ln Y^2$$
$$+ \sum_i a_i \ln P_i$$
$$+ 1/2 \sum_i \sum_j \beta_{ij} \ln P_i \ln P_j$$
$$+ 1/2 \beta_{iy} \ln Y \ln P_i \quad (4:2)$$

where $i,j = K,L$

where $\ln C^*$ is the logarithm of total costs

$\ln P_i$ is the logarithm of the ith input price

$\ln Y$ is the logarithm of output

By Shephards's Lemma, factor shares are derived by differentiating equation (4:2) with respect to each logged input price. Then

$$\delta \ln C / \delta \ln P_i = P_i X_i / C^* = S_i$$

and the input demand functions, in terms of cost shares, take the form

$$S_i = a_i + \sum_j \beta_{ij} \ln P_j + \beta_{iy} \ln Y \quad (4:3)$$

where $i,j = K,L$.

In order for the translog to satisfy linear homogeniety in input prices, and other properties of a well behaved production function, the following parameter restrictions are required.

$\sum_i a_i = 1$ [linear homogeniety]

$\sum_i \beta_{ij} = 0$ [cournot aggregation]

$\sum_i \beta_{iy} = 0$ [Engel aggregation]

$\beta_{ij} = \beta_{ji}$ [sluksky symmetry]

The appearance of output in the translog cost function in equation (4:2) introduces nonhomotheticity. Thus, by restricting $\beta_{iy} = 0$ produces the linear expansion path of the homothetic case. Further restricting θ to equal zero produces a technology which is homogeneous of degree $1/\beta$. Setting $\beta = 1$, would then yeild a cost function homogeneous of degree one.

The popularity of the translog cost function is due to the ease of estimation. Furthermore, the translog cost function allows arbitrary configurations of the matrix of elasticities of substitution. It also permits variations in these elasticities across input pairs or prices. Unlike elasticities derived from Cobb-Douglas or CES production functions, the translog cost function permits complementarities.

Mundlak (1968) provides three alternative measures of substitutability between pairs of inputs X_i, X_j. When the elasticity of substitution is positive inputs are considered as substitutes and if the elasticity of substitution is negative, inputs are complements. Uzawa (1962) has measured the Allen partial elasticity of substitution between inputs i & j as

$$\sigma_{ij}^A = \frac{C^* \cdot \delta^2 C^*/\delta\rho i\, \delta\rho j}{\delta C^*/\delta\rho_i \cdot \delta C^*/\delta\rho j} \qquad (4:4)$$

Berndt-Wood (1975) show that for the translog cost function, the Allen partial elasticity of substitution is

$$\sigma_{ij}^A = \frac{\beta_{ij} + S_i S_j}{S_i S_j} \qquad (4:4a)$$

$$\sigma_{ii}^A = \frac{\beta_{ij} + S_i^2 - S_i}{S_i^2} \qquad (4:4b)$$

where σ_{ij} = cross partial elasticity of substitution between inputs i, j

σ_{ii} = own partial elasticity of substitution between inputs i,j

The Morishima elasticity of substitution is written by Koizimi (1976) as

$$\sigma_{ij}^M = S_j (\sigma_{ij}^A - \sigma_{jj}^A) \qquad (4:5a)$$

and the shadow elasticity of substitution by McFadden (1963) is

$$\sigma_{ij}^S = \frac{S_i S_j (2\sigma_{ij}^A - \sigma_{ii}^A - \sigma_{jj}^A)}{S_i + S_j} \quad (4:5b)$$

Further, it has been shown by Binswanger (1974) that the elasticity of substitution between pairs of inputs can be calculated as

$$\sigma_{ij}^B = \beta_{ij}/S_i S_j + 1 \quad (4:6a)$$

$$\sigma_{ii}^B = \beta_{ii}/S_i - 1/S_i + 1 \quad (4:6b)$$

Assuming Hicks neutral technical change (homotheticity) and constant returns to scale (CRTS), the translog cost function can be rewriten as

$$\ln C(P_i, P_j) = a_o + \sum_i a_i \ln P_i$$
$$+ 1/2 \sum_i \sum_j \beta_{ij} \ln P_i \ln P_j \quad (4:7)$$

and the factor demands in terms of cost shares are,

$$S_i = a_i + \sum_j \beta_{ij} \ln P_j \quad (4:8)$$

where $i,j = K,L$

alternatively

$$S_K = a_K + \beta_{KK} \ln P_K + \beta_{KL} \ln P_L \quad (4:8a)$$

$$S_L = a_L + \beta_{LK} \ln P_K + \beta_{LL} \ln P_L \quad (4:8b)$$

In order that the system of cost share equations [4:8a] and [4:8b] satisfy the regularity conditions and the properties of the neo-classical production function, the following restrictions are required.

$a_K + a_L = 1$ [linear homogeniety]

$\left.\begin{array}{l}\beta_{KL} + \beta_{KK} = 0 \\ \beta_{KL} + \beta_{LL} = 0\end{array}\right\}$ [cournot aggregation]

$$\beta_{KL} = \beta_{LK} \qquad \text{[sluksky symmetry]}$$

To see the effects of restrictions on the estimating form of the cost share equations, the following restrictions are imposed on the cost share equations as follows; by linear homogeneity,

$$a_K + a_L = 1,$$

i.e. $a_K = 1 - a_L$ or $a_L = 1 - a_K$ \hfill (4:9)

By symmetry,

$$\beta_{KL} = \beta_{LK}$$

and by cournot aggregation, i.e. the row and column sum of the $_{ij}$ matrix is zero.

$$\beta_{KK} + \beta_{KL} = 0 \qquad (4:10)$$

$$\beta_{LK} + \beta_{LL} = 0 \qquad (4:11)$$

Given the cost share equations [4:8a] and [4:8b) and using [4:10] and [4:11], gives

$$\beta_{KK} = -\beta_{KL}$$

$$\beta_{LL} = -\beta_{LK}$$

Therefore one equation is redundant. Thus, with these regularity restrictions, the cost share equation to be estimated is

$$S_K = a_K + \beta_{KK} \ln P_K + \beta_{KL} \ln P_L \qquad (4:12a)$$

or $\quad S_L = a_L + \beta_{LK} \ln P_K + \beta_{LL} \ln P_L \qquad (4:12b)$

The estimates for a_L, β_{LK} and β_{LL} can be calculated ex-post by substituting the parameter estimates into equations [4:9], [4:10] and [4:11]. Since the set of simultaneous equations have been reduced to a single linear equation, equation [4:12] can be estimated using OLS. In this two-factor case, therefore, there is no computational problems related to which equation is ommited.

Further conditions of a well-behaved production function are that output should increase monotonically with all inputs and that the isoquants are convex. The translog does not satisfy these restrictions globally. In fact, when at least one $\beta_{ij} = 0$, there exist configurations of inputs such that neither monotonicity nor convexity is satisfied. This follows simply from the quadratic nature of the translog function. On the other hand, there are regions in the input space where these conditions are satisfied. For any set of parameters and input levels the monotonicity and convexity conditions can be easily checked. Monotonicity requires that $\delta F/\delta X_i > 0$. Since F and X_i are always positive, an equivalent set of conditions is that the cost share equations are positive. Assuming markets are competitive, the set of necessary conditions for efficient production is that $\delta F/\delta X_i = R_i$ where R_i is the price of ith input. Then monotonicity conditions can be written as

$$S_i = \frac{\delta \ln F}{\delta \ln X_i} = \frac{\delta F}{\delta X_i} \cdot \frac{X_i}{F} = \frac{P_i X_i}{F} > 0$$

The isoquants of the translog function are strictly convex if the corresponding bordered Hessian matrix of first and second partial derivatives is negative definite. This can be evaluated at each data point for any estimated translog function.

Measurement of technical change via the translog cost function approach

Assuming homotheticity and constant returns to scale, the translog cost function is

$$\ln C(P_i, P_j) = a_0 + \sum_i a_i \ln P_i + 1/2 \sum_i \sum_j \beta_{ij} \ln P_i \ln P_j \quad (4:7)$$

The above funtion assumes Hick's-neutral technical change. A further property of the translog cost function is that it permits respecification of the estimation equation to include the effects of factor - augmented technical change. As shown by Berndt and Khaled (1979) and Lopez (1980), variations in factor shares can be explained not only by relative prices but by technical change by relating the dependent variables to time. The translog cost function becomes

$$\ln C^* = a_o + \sum_i a_i \ln P_i + 1/2 \sum_i \sum_j \beta_{ij} \ln P_i \ln P_j$$

$$+ \beta_T \cdot T + \sum_i \beta_{iT} \ln P_i \cdot T$$

$$+ 1/2 \beta_{TT} \cdot T^2 \qquad (4:13)$$

Differentiating equation [4:13] with respect to log P_i and invoking Shephard's Lemma yeilds the factor demand equations which are expressed in terms of factor cost shares with an additional variable time (T).

$$S_i = a_i + \sum_i \beta_{ij} \ln P_j + \sum_i \beta_{iT} T \qquad (4:14)$$

Examples of respecifying the translog cost function to permit technical change by adding the independent variable time (T) include Berndt & Wood (1975), Binswanger (1974), Ray (1980) and Ball and Chambers (1982).

Data, Estimation and Results of the Translog Cost Function for the Malaysian Manufacturing Industries

This section consists of discussion of the data and related problems concerning the operational definition of variables for the translog cost estimation, the estimation procedures and the discussion of results of the translog cost function.Since the CES-translog cost function is only a variant of the CES and translog cost function, the discussion of data and estimation is also discussed in this section.

Sources of data and measurement of variables

To ensure comparability of alternative estimates of the elasticity of substitution in 5-digit Malaysian manufacturing industries, all data for the estimation of the translog cost function and the CES-translog cost function are from the Surveys/Censuses of Manufacturing Industries, West Malaysia, the Industrial Surveys of Malaysia and Manufacturing Division, Department of Statistics, Malaysia. However, data for value of fixed assets and value of depreciation are available for only a number of years beginning in 1969. As such, the estimation of the translog cost function and the CES-translog cost function are based on a time-series from 1969 to 1984 for fifty 5-digit Malaysian manufacturing industries.

The chief sources of data are the Surveys/Censuses of Manufacturing Industries which have been conducted annually up to 1976. The Surveys/Censuses are followed by the Industrial Surveys of Malaysia from 1978 to 1984. The information in the Surveys and Censuses includes value - added (VA), number of workers employed (L), wages and salaries (W), cost of inputs (C), and value of fixed assets (FA). There were no surveys for 1977 and 1980. Furthermore, the information on value of fixed assets are not consistently given for the whole period 1970-1984.

Information on value of depreciation (D), value of fixed assets (FA) and the breakdown for cost of inputs are derived from the Manufacturing Division, Department of Statistics, Malaysia. Consumer Price Index (CP1) and Industrial Production Index (IPI) are taken from Bank Negara Annual Report, 1985.

The final data required for the estimation of the translog cost function and the CES-translog cost function are the cost share of capital (S_K), the cost share of labor (S_L), the service price of capital (P_K) and the service price of labor (P_L), and the first derivative of the logarithm of total cost with respect to time (S_T).

The estimation of the translog cost function and the CES-translog cost function for Malaysia would be most befitting if data on cost shares and service prices of capital and labor can be constructed following procedures outlined by Christensen-Jorgenson (1969, 1970) and Berndt & Christensen (1970). Such procedures however would require extra information on variations of effective tax rates, rates of return, capital gains, years of education of labor force and others in order to construct the Divisia quality indexes for capital and labor.

The estimation in this study is however based on less sophisticated procedures of constructing the final data for S_K, S_L, S_T, P_K and P_L.

Following Wills (1979) and Vashist (1985), the cost share of capital and labor are constructed as follows,

$$S_K = (VA - S_L)/TC$$

$$S_L = W/TC$$

and following procedures by Ioannides and Caramanis (1979), Wills (1979) and Vashist (1985), the service prices of capital and labor, P_K and P_L are constructed as follows,

$$P_K = S_K/K$$

$$P_K = S_L/L$$

where VA = value-added

L = number of full-time workers plus half of part-timeworkers

W = total wages and salaries

TC = total cost to industry measured as total cost of inputs + fixed cost + wages and salaries

K   value of fixed assets + value of circulating capital (materials + electricity, fuel, lubricants and water + intermediate supplies).

Initially it was intended to construct the capital stock series by the well-known perpetual inventory method (Christensen & Jorgenson, 1969). However, there is no benchmark available and there is also serious deficiency of information on the age structure of existing capital stock.

Another problem in the measurement of capital is related to the computation of total cost and hence the service prices of capital and labor. Total cost includes cost of circulating capital and fixed cost as the cost of capital while the wage bill is the cost of labor. Thus, despite various limitations, such as reconciling stock and flow concepts, the value of fixed assets and value of circulating capital is taken to represent our capital data, K. Another reason is that the objective of the research is to calculate the substitution possibilities between capital and labor only. As such capital should be the residual of labor inputs (Fuss, 1977).

Estimation of the translog cost function

Assume that there exists in Malaysian manufacturing sector a twice differentiable aggregate production function relating the flow of gross output to the services of capital and labor. Further, assume that production is characterised by constant returns to scale and that any technical change affecting capital and labor is Hicks-nuetral. For purposes of estimation, the set of simultaneous equations (4:8a) and (4:8b) can be used. However, data from Surveys and Censuses of Manufacturing Companies, West Malaysia and the Industrial Surveys of Malaysia are subject to errors. These errors can result in deviations of the actual cost shares from the cost minimising shares. Kulatilaka (1985) shows that stochastic specifications introduce additive errors due to errors in measurement of output and cost shares. These errors are specified in the error term (e_i) and the estimating equation takes a stochastic form

$$S_i = a_i + \sum_i \beta_{ij} \ln P_j + e_i \qquad (4:15)$$

In a two-factor case, the estimating equations are sets of simultaneous equations, in the following stochastic form,

$$S_K = a_K + \beta_{KK} \ln P_K + \beta_{KL} \ln P_L + e_K \qquad (4:16)$$

$$S_L = a_L + \beta_{LK} \ln P_K + \beta_{LL} \ln P_L + e_L \qquad (4:17)$$

The parameters of the translog cost function in Malaysian manufacturing sector can be estimated using equations (4:16) and (4:17). For the system of share equations (4:16) and (4:17), the disturbances are likely to be correlated across equations. Therefore e_K and e_L will be correlated. This suggests that the Iterative Zellner Efficient Method or Seemingly Unrelated Regression (SUR) method will give efficient parameter estimates. Zellner (1962) has shown that when disturbances across equations are correlated, and if the correlation is known, then the parameters can be estimated more efficiently by taking this information into account. Furthermore, Zellner (1963) has demonstrated that even when the correlation is unknown, it is likely that using an estimate of the correlation will improve estimation efficiency.

Firstly, alternative versions of Model (B1) to Model (B6) are estimated without restrictions using the Seemingly Unrelated Regression method from time-series data 1969 - 1984 in 5-digit Malaysian manufacturing industries.

[MODEL B1] $\quad S_K \quad = \quad a_K + \beta_{KK} \ln P_K + \beta_{KL} \ln P_L + \epsilon_K$

$\qquad\qquad\quad S_L \quad = \quad a_L + \beta_{LK} \ln P_{PK} + \beta_{LL} \ln P_L + \epsilon_L$

[MODEL B2] $\quad S_K(AR1) \quad = \quad a_K + \beta_{KK} \ln P_K + \beta_{KL} \ln P_L + \epsilon_K$

$\qquad\qquad\quad S_L(AR1) \quad = \quad a_L + \beta_{LK} \ln P_K + \beta_{LL} \ln P_L + \epsilon_L$

[MODEL B3] $\quad SKR \quad = \quad a_K + \beta_{KK} \ln P_{KR} + \beta_{LK} \ln P_{LR} + \epsilon_K$

$\qquad\qquad\quad SLR \quad = \quad a_L + \beta_{LK} \ln P_{KR} + \beta_{LL} \ln P_{LR} + \epsilon_L$

[MODEL B4] $\quad SKR(AR1) \quad = \quad a_K + \beta_{KK} \ln P_{KR} + \beta_{KL} \ln P_{LR} + \epsilon_K$

$\qquad\qquad\quad SLR(AR1) \quad = \quad a_L + \beta_{LK} \ln P_{KR} + \beta_{LL} \ln P_{LR} + \epsilon_L$

[MODEL B5] SK1 $= a_K + \beta_{KK} \ln P_{K1} + \beta_{KL} \ln P_{L1} + \epsilon_K$

SL1 $= a_L + \beta_{LK} \ln P_{K1} + \beta_{LL} \ln P_{L1} + \epsilon_L$

[MODEL B6] SKI(AR1) $= a_K + \beta_{KK} \ln P_{KI} + \beta_{KL} \ln P_{LI} + \epsilon_K$

SLI(AR1) $= a_L + \beta_{LK} \ln P_{KI} + \beta_{LL} \ln P_{LI} + \epsilon_L$

where

S_K is cost share of capital

S_L is cost share of labor

P_K is service price of capital

P_L is service price of labor

$S_K(AR1)$ is cost share of capital corrected for auto-correlation

$S_L(AR1)$ is cost share of labor corrected for auto-correlation

SKR is real cost share of capital with CPI as deflator, 1980 = 100

SLR is real cost share of labor with CPI as deflator, 1980 = 100

SKR(AR1) is real cost share of capital (CP1) corrected for auto-correlation

SLR(AR1) is real cost share of labor (CP1) corrected for auto-correlation

SK1 is real cost share of capital with Industrial Production Index (IPI) as deflator, 1980 = 100

SL1 is real cost share of labor

Next, the equations (4:16) and (4:17) are estimated with restrictions.

A number of methods can be used, including Zellners' generalised least squares (GLS) estimation procedure which yeilds estimators which are sensitive to which cost share equation is deleted from the system of equations. A maximum likelihood procedure would provide parameter estimates which are invariant to the choice of equations to be actually estimated (Barken, 1969). However, Kmenta and Gilbert (1968) have demonstrated that Full Information Maximum Likelihood (FIML) and Iterated Zellner Efficient Estimation (IZEF), commonly known as Seemingly Unrelated Regression (SUR), lead to identical estimates. Rubble (1968) has also shown the computational equivalence of IZEF and FIML estimators.

However, in order to use FIML procedure, input prices (P_K, P_L) and output must be exogeneous and, thus, orthogonal to the additive errors (Kulatilaka, 1985). If the data are for a complete economy then output and factor prices are likely to be endogeneous. Furthermore, as demonstrated by Berndt and Savin (1975), neither the maximum likelihood estimators nor the Zellner's estimators will be invariant to the equation deleted if the error terms in the model are not well behaved (i.e. presence of autocorrelation).

In this case of the two-input translog function, one equation is redundant and can be omitted. Basing on Models [B1] - [B6] both cost share equations S_K and S_L are estimated separately using OLS procedures. In the two-input case, it does not matter which equation is deleted. The final choice of the estimating equation will be based on the statistical results.

The third procedure is to determine the effect of techical change. The estimating equations to measure the effect of technical change in the Malaysian manufacturing sector are as follows;

$$S_K = a_K + \beta_{KK} \ln P_K + \beta_{KL} \ln P_L + \beta_{KK} T + \beta_{KK} T \qquad (4:18)$$

$$S_L = a_L + \beta_{LK} \ln P_K + \beta_{LL} \ln P_L + \beta_{KL} T + \beta_{LL} T \qquad (4:19)$$

$$S_T = a_T + \beta_{KT} \ln P_K + \beta_{LT} \ln P_L + \beta_{TT} T \qquad (4:20)$$

where $\quad S_T = \dfrac{\delta \ln C^*}{\delta \ln T}$

The terms (β_{LT} and β_{KT}) are estimates of the factor-saving Hicks biases of technical change, since they measure the rate of change in the cost shares not attributable to prices.

Thus,

if $\beta_{iT} = 0$ implies Hicks neutral technical change

  $\beta_{LT} > 0$ implies labor-saving technical change

  $\beta_{KT} > 0$ implies capital-saving technical change

through a series of simple manipulation equations [4:18] [4:19] and [4:20] can be rewritten as follows:

$$S_K = a_K + \beta_{KK}[\ln P_k + T] + \beta_{KL}[\ln P_L + T] \qquad (4:21)$$

$$S_L = a_L + \beta_{KL}[\ln P_K + T] + \beta_{LL}[\ln P_L + T] \qquad (4:22)$$

The final estimating equations to measure the effect of technical change in the Malaysian manufacturing sector are the set of simultaneous equations (4:23) (4:24) and (4:25). These equations are deterministic. However errors of optimisation can arise due to deviations of the firm's actual behaviour from it's cost-minimising behaviour. To account for such random errors, disturbance terms were added to the equations so that they take the following set of stochastic simultaneous equations

$$S_K = a_K + \beta_{KK}[\ln P_K + T] + \beta_{KL}[\ln P_L + T] + \epsilon_K \qquad (4:23)$$

$$S_L = a_L + \beta_{LK}[\ln P_K + T] + \beta_{LL}[\ln P_L + T] + \epsilon_L \qquad (4:24)$$

$$S_T = a_T + \beta_{KT} \ln P_K + \beta_{LT} \ln P_L + \beta_{TT} T \qquad (4:25)$$

Based on these equations, alternative regression models [B7] to [B12] are estimated using the Seemingly Unrelated Regression procedure.

MODEL [B7] $S_K = a_K + \beta_{KK} \ln PKT + \beta_{KL} \ln PLT + \epsilon_k$

  $S_L = a_L + \beta_{LK} \ln PKT + \beta_{KL} \ln PLT + \epsilon_L$

  $S_T = a_T + \beta_{KT} \ln PK + \beta_{LT} \ln PL + \beta_{TT} T$

Translog Cost Function 63

MODEL [B8] $S_K(AR1) = a_K + \beta_{KK} \ln PKT + \beta_{KL} \ln PLT + \epsilon_k$

$S_L(AR1) = a_L + \beta_{LK} \ln PKT + \beta_{LL} \ln PLT + \epsilon_L$

$S_T = a_T + \beta_{KT} \ln PK + \beta_{LT} \ln PL + \epsilon_{TT} T$

MODEL [B9] $SKR = a_K + \beta_{KK} \ln PKRT + \beta_{KL} \ln PLRT + \epsilon_K$

$SLR = a_L + \beta_{LK} \ln PKRT + \beta_{LL} \ln PLRT + \epsilon_k$

MODEL [B10] $SKR(AR1) = a_K + \beta_{KK} \ln PKRT + \beta_{KL} \ln PLRT + \epsilon_K$

$SLR(AR1) = a_L + \beta_{LK} \ln PKRT + \beta_{LL} \ln PLT + \beta_{TT} T$

$CTR = a_T + \beta_{LT} \ln PKT + \beta_{LL} \ln PLT + \beta_{TT} T$

MODEL [B11] $SKI = a_K + \beta_{KK} \ln PKIT + \beta_{KL} \ln PLIT + \epsilon_K$

$SLI = a_L + \beta_{LK} \ln PKIT + \beta_{LL} \ln PLIT + \epsilon_K$

$CTI = a_T + \beta_{LT} \ln PKT + \beta_{LL} \ln PLT + \beta_{TT} T$

MODEL [B12] $SKI(AR1) = a_K + \beta_{KK} \ln PKIT + \beta_{KL} \ln PLIT + \epsilon_K$

$SLI(AR1) = a_L + \beta_{LK} \ln PKIT + \beta_{LL} \ln PLIT + \epsilon_L$

$CTI = a_T + \beta_{LT} \ln PKT + \beta_{LL} \ln PLT + \beta_{TT} T$

ln PKT   is the logarithm of service price of capital plus time; 1969 = 1

ln PLT   is the logarithm of service price of labor plus time; 1969 = 1

S_T is the lagged values of total cost

SKR   is the real cost share of capital deflated with CPI, 1980 = 100

SLR   is the real cost share of labor deflated with CPI, 1980 = 100

ln PKRT   is the logarithm of real service price of capital deflated by CPI

ln PLRT   is the logarithm of real service price of labor deflated by CPI

SKR(AR1)   is real cost share of capital (CPI) corrected for auto-correlation

SLR(AR1)   is real cost share of labor (CPI) corrected for auto-correlation

SKI   is real cost share of capital deflated by Industrial Production Index (IPI), 1980 = 100

SLI   is real cost share of labor deflated by the Industrial Production Index (CPI), 1980 = 100

ln PKIT   is the logarithm of the service price of capital plus time

ln PLIT   is the logarithm of the service price of labor plus time

CTR   is the lagged values of real total cost deflated by CPI

CTI   is the lagged values of real total cost deflated by IPI

The final choice of the model to estimate the elasticity of substitution with non-neutral technical change, is based on various statistical results.

Discussion of empirical results

Elasticity of substitution without technical change.

The most appropriate equation to estimate the elasticity of substitution between capital and labor in the Malaysian manufacturing sector is chosen based on both statistical and theoretical reasons. A statistical search for the best fitting equation based on parameter estimates, the conventional R^2 and the Durbin - Watson statistic, is carried out for 50 5-digit industry groups. As shown in Table 4.1, in 25 cases, Model [B4] has the best fit while in the other half of the 5-digit industry groups, Model [B6] is the best fitting equation. The differences of the results of these two models however are very small. For example, for Pineapple

Table 4.1. Statistical performance of alternative models of the unrestricted translog cost function estimation of Malaysian manufacturing industries, 1969-1984

INDUSTRY\MODEL			[B1]	[B2]	[B3]	[B4]	[B5]	[B6]
Slaughtering Preparing & Preserving Meat	SK:[a]	R^2	.34	.90	.63	.93	.70	.93
		DW[b]	.98	1.1	1.0	1.5	.99	1.6
	SL:[c]	R^2	.15	.32	.39	.50	.49	.56
		DW	1.3	1.6	1.4	1.5	1.4	1.5
Ice Cream Manufacturing	SK:	R^2	.72	.83	.93	.78	.82	.77
		DW	1.3	1.4	1.3	2.0	1.1	2.1
	SL:	R^2	.66	.74	.82	.78	.82	.77
		DW	1.1	2.1	.80	1.9	.81	1.9
Manufacture of Other Dairy Products	SK:	R^2	.17	.11	.67	.70	.73	.77
		DW	2.0	2.0	2.5	2.2	2.4	2.1
	SL:	R^2	.62	.95	.74	.77	.75	.79
		DW	2.0	1.5	2.3	2.0	2.3	2.1
Pineapple Canning	SK:	R^2	.68	.88	.96	.96	.96	.96
		DW	.55	1.4	1.4	1.8	1.3	1.6
	SL:	R^2	.70	.87	.90	.84	.94	.83
		DW	.70	1.2	1.1	1.2	1.3	1.2
Other Canning and Preserving of Fruits and Vegetables	SK:	R^2	.45	.44	.61	.51	.64	.60
		DW	.65	1.8	.80	1.6	1.6	1.8
	SL:	R^2	.60	.54	.82	.87	.83	.83
		DW	1.6	2.0	.80	2.0	1.7	1.9
Coconut Oil Manufacturing	SK:	R^2	.90	.92	.77	.86	.77	.86
		DW	1.1	1.8	1.0	1.8	1.3	1.9
Palm Oil Manufacturing	SK:	R^2	.67	.74	.91	.91	.42	.91
		DW	2.2	2.1	1.7	1.7	1.1	1.7
	SL:	R^2	.07	.14	.21	.03	.26	.03
		DW	2.5	2.1	2.4	2.0	1.4	2.0
Palm Kernel Oil Manufacturing	SK:	R^2	.64	.92	.86	.93	.85	.89
		DW	1.6	.71	2.1	2.0	1.7	.99
	SL:	R^2	.06	.60	.49	.84	.59	.86
		DW	.71	.90	.92	.92	1.3	1.1
Vegetable and Animal Oils & Fats	SK:	R^2	.54	.50	.79	.78	.80	.78
		DW	1.6	1.9	1.9	2.0	1.8	2.0
	SL:	R^2	.86	.82	.95	.92	.96	.94
		DW	1.5	1.7	1.3	1.6	1.2	1.6
Rice Milling	SK:	R^2	.41	.49	.82	.83	.59	.50
		DW	1.8	1.7	1.7	1.6	1.2	1.7
	SL:	R^2	.48	.68	.83	.72	.63	.77
		DW	.74	1.6	1.6	1.8	1.1	1.8
Biscuit Factories	SK:	R^2	.03	.53	.93	.95	.90	.91
		DW	2.1	1.6	.75	2.0	.75	2.0

Table 4.1. Continued

INDUSTRY\MODEL			[B1]	[B2]	[B3]	[B4]	[B5]	[B6]
	SL:	R^2	.23	.48	.91	.93	.91	.92
		DW	1.7	2.0	.61	1.6	.47	1.5
Sugar Factories and Refineries	SK:	R^2	.72	.83	.89	.93	.89	.94
		DW	1.2	1.9	2.2	1.2	2.4	1.1
	SL:	R^2	.95	.96	.97	.97	.98	.98
		DW	1.2	1.9	1.4	1.9	1.7	1.9
Manufacture of Cocoa, Chocolate and Sugar Confectionery	SK:	R^2	.54	.80	.87	.86	.86	.85
		DW	1.4	2.1	1.6	1.9	1.7	1.9
	SL:	R^2	.35	.78	.92	.93	.94	.93
		DW	1.1	1.7	1.0	1.8	1.1	1.8
Ice Factories	SK:	R^2	.79	.91	.84	.89	.82	.85
		DW	.95	1.4	.78	1.8	.78	1.8
	SL:	R^2	.45	.82	.76	.84	.76	.83
		DW	.83	1.6	.78	1.9	.80	1.9
Coffee Factories	SK:	R^2	.001	.77	.83	.89	.90	.91
		DW	.89	1.0	1.1	1.8	1.0	2.0
	SL:	R^2	.07	.53	.83	.85	.88	.88
		DW	1.2	2.1	1.5	2.0	1.3	2.1
Meehoon, Noodles & Related Products	SK:	R^2	.006	.70	.87	.91	.89	.91
		DW	1.8	2.2	1.5	2.2	1.6	2.2
	SL:	R^2	.82	.83	.90	.89	.90	.89
		DW	1.5	1.9	1.5	1.9	1.4	1.9
Manufacture of Prepared Animal Feeds	SK:	R^2	.05	.14	.92	.93	.94	.92
		DW	1.8	1.9	1.1	1.7	1.1	1.4
	SL:	R^2	.33	.75	.90	.98	.94	.94
		DW	.95	1.9	1.1	2.0	1.2	2.1
Soft Drinks and Carbonated Beverages	SK:	R^2	.46	.82	.90	.89	.91	.90
		DW	.69	1.7	.65	1.3	.48	1.3
	SL:	R^2	.62	.74	.87	.87	.88	.86
		DW	1.2	1.9	.62	1.6	5.2	1.5
Tobacco Manufacturing	SK:	R^2	.14	.93	.64	.92	.76	.93
		DW	1.1	.84	1.8	1.1	1.6	1.0
	SL:	R^2	.10	.83	.53	.54	.68	.68
		DW	1.2	1.3	2.0	2.0	2.0	2.0
Manufacture of Leather & Leather Products	SK:	R^2	.72	.72	.78	.83	.78	.82
		DW	.80	1.6	1.3	1.9	1.5	1.8
	SL:	R^2	.42	.36	.32	.32	.35	.49
		DW	1.6	1.6	1.8	1.8	2.3	2.1
Sawmilling	SK:	R^2	.79	.87	.85	.85	.91	.89
		DW	1.5	1.2	1.6	1.3	1.6	1.7

Table 4.1. Continued

INDUSTRY\MODEL			[B1]	[B2]	[B3]	[B4]	[B5]	[B6]
	SL:	R^2	.73	.73	.88	.87	.92	.91
		DW	2.0	1.9	1.6	1.9	1.5	1.8
Planning Mills & Joinery Works	SK:	R^2	.50	.44	.89	.91	.92	.96
		DW	2.0	1.9	2.1	2.0	2.5	1.9
	SL:	R^2	.82	.89	.94	.94	.94	.92
		DW	.95	1.4	.67	1.5	.76	1.7
Manufacture of Furniture & Fixtures	SK:	R^2	.16	.55	.92	.93	.93	.93
		DW	1.6	1.8	1.2	1.8	1.2	1.7
	SL:	R^2	.79	.89	.91	.92	.91	.91
		DW	1.5	1.8	1.2	2.0	1.2	2.0
Clothing Factories	SK:	R^2	.77	.96	.92	.98	.90	.97
		DW	1.1	.99	.98	1.3	.89	1.3
	SL:	R^2	.31	.67	.86	.88	.87	.88
		DW	1.4	1.7	1.5	2.3	1.3	2.1
Manufacture of Paper & Paper Products	SK:	R^2	.47	.48	.65	.67	.65	.72
		DW	2.0	2.0	2.1	2.0	2.1	2.0
	SL:	R^2	.003	.41	.79	.89	.81	.90
		DW	1.5	1.8	2.2	1.2	2.2	1.4
Printing Publishing & Allied Industries	SK:	R^2	.80	.95	.92	.95	.91	.97
		DW	.78	1.9	1.3	2.0	1.6	1.2
	SL:	R^2	.15	.72	.92	.93	.87	.92
		DW	1.0	2.1	.89	1.8	.76	1.4
Manufacture of Basic Industrial Chemicals	SK:	R^2	.83	.84	.89	.94	.88	.93
		DW	1.9	1.9	.72	1.3	.81	1.2
	SL:	R^2	.16	.60	.82	.90	.87	.91
		DW	.46	1.1	.78	1.3	.68	1.2
Manufacture of Chemical Fertiliser & Pesticides	SK:	R^2	.37	.61	.86	.83	.91	.91
		DW	.81	1.5	1.5	2.1	1.9	2.0
	SL:	R^2	.91	.98	.91	.98	.90	.97
		DW	.65	1.3	.71	1.4	.73	1.4
Manufacture of Paints, Varnishes and Lacquers	SK:	R^2	.44	.50	.96	.96	.98	.97
		DW	2.0	2.0	1.8	1.9	1.6	1.8
	SL:	R^2	.76	.81	.97	.96	.97	.96
		DW	1.3	1.6	1.1	1.5	1.2	1.7
Manufacture of Drugs, Medicine & Pharmacueticals	SK:	R^2	.59	.71	.86	.84	.90	.88
		DW	1.1	2.1	1.3	2.1	.73	1.4
	SL:	R^2	.74	.82	.91	.91	.79	.55
		DW	1.0	1.6	1.1	1.7	1.1	1.7
Manufacture of Soap & Cleaning Preparation	SK:	R^2	.54	.76	.92	.89	.96	.94
		DW	1.0	1.7	1.3	1.7	1.2	1.7
	SL:	R^2	.93	.96	.97	.97	.96	.95
		DW	.90	1.7	.83	2.1	.84	2.1

Table 4.1. Continued

INDUSTRY\MODEL			[B1]	[B2]	[B3]	[B4]	[B5]	[B6]
Manufacture of Perfumes, Cosmetics & Toileteries	SK:	R^2	.46	.52	.65	.65	.67	.64
		DW	.71	1.8	.43	1.3	.54	1.6
	SL:	R^2	.28	.48	.83	.82	.83	.82
		DW	2.5	2.1	1.4	2.0	1.2	2.0
Petroleum Refineries	SK:	R^2	.16	.29	.57	.77	.68	.82
		DW	.72	1.1	.74	1.1	.73	1.2
	SL:	R^2	.43	.84	.71	.86	.73	.86
		DW	.45	.49	.64	.56	.64	.61
Petroleum & Coal Products	SK:	R^2	.93	.93	.93	.93	.93	.93
		DW	1.7	1.8	1.8	1.8	1.7	1.7
	SL:	R^2	.91	.91	.92	.92	.92	.92
		DW	1.7	1.8	1.8	1.8	1.7	1.8
Rubber Products	SK:	R^2	.30	.29	.16	.19	.51	.47
		DW	2.0	2.0	1.1	2.0	1.9	1.9
	SL:	R^2	.24	.24	.17	.21	.42	.27
		DW	1.8	1.9	.93	2.0	1.5	2.1
Plastic Products	SK:	R^2	.48	.83	.91	.93	.92	.93
		DW	.92	1.7	.43	1.1	.42	1.3
	SL:	R^2	.39	.83	.85	.89	.86	.89
		DW	.92	1.7	.78	1.9	.84	1.9
Pottery China and Earthernware	SK:	R^2	.69	.73	.77	.77	.74	.76
		DW	1.5	1.8	1.6	1.8	1.5	1.8
	SL:	R^2	.76	.77	.89	.87	.90	.88
		DW	1.6	1.9	1.5	1.7	1.3	1.6
Hydraulic Cement	SK:	R^2	.72	.69	.87	.80	.89	.82
		DW	.77	2.1	.85	1.9	.71	1.9
	SL:	R^2	.70	.79	.95	.93	.95	.94
		DW	1.1	1.7	.81	1.6	.79	2.0
Cement & Concrete	SK:	R^2	.64	.82	.93	.90	.92	.91
		DW	.87	1.6	.91	1.4	.64	1.5
	SL:	R^2	.61	.83	.89	.96	.89	.95
		DW	.91	1.9	1.6	1.1	1.7	1.2
Iron Foundaries	SK:	R^2	.80	.85	.96	.95	.97	.98
		DW	1.0	1.6	1.5	1.7	2.4	2.0
	SL:	R^2	.90	.76	.88	.83	.90	.88
		DW	1.2	1.5	1.1	1.3	1.1	1.4
Non-Ferrous Metal Product	SK:	R^2	.88	.98	.90	.92	.86	.85
		DW	3.3	1.6	2.2	1.3	1.5	1.7
	SL:	R^2	.18	.19	.83	.86	.88	.91
		DW	1.9	1.9	2.2	1.9	2.4	1.9
Wire Products Manufacturing	SK:	R^2	.60	.83	.72	.74	.75	.74
		DW	1.6	1.2	2.0	1.6	1.96	1.6

Table 4.1. Continued

INDUSTRY\MODEL			[B1]	[B2]	[B3]	[B4]	[B5]	[B6]
	SL:	R^2	.58	.57	.88	.87	.91	.87
		DW	1.9	2.0	1.5	1.8	1.5	1.8
Brass, Copper Pewter & Alluminium Product	SK:	R^2	.54	.64	.95	.95	.96	.96
		DW	1.5	1.7	2.1	2.1	1.9	1.9
	SL:	R^2	.57	.74	.95	.93	.97	.96
		DW	.89	1.5	1.1	1.5	1.1	1.9
Industrial Machinery & Parts	SK:	R^2	.54	.64	.95	.95	.96	.96
		DW	2.4	2.2	2.1	2.0	1.9	1.9
	SL:	R^2	.86	.87	.94	.88	.94	.88
		DW	1.2	1.6	1.3	1.4	1.4	1.4
Electrical Machinery, Approatus & Appliances	SK:	R^2	.16	.27	.84	.89	.86	.91
		DW	2.7	2.2	2.5	2.1	2.5	1.9
	SL:	R^2	.08	.03	.88	.90	.92	.93
		DW	1.8	1.7	1.5	1.9	1.8	1.6
Shipbuilding Boatmaking & Repairing	SK:	R^2	.59	.71	.81	.83	.80	.81
		DW	1.1	2.2	2.2	2.0	2.1	2.0
	SL:	R^2	.74	.82	.90	.88	.92	.90
		DW	1.0	1.6	1.5	1.9	1.0	1.9
Manufacturing of Motor Vehicle Bodies	SK:	R^2	.18	.67	.92	.93	.91	.91
		DW	1.5	1.9	1.4	1.7	1.0	2.1
	SL:	R^2	.48	.79	.87	.86	.85	.83
		DW	1.2	2.0	.85	2.0	.81	1.8
Manufacture of Motor Vehicle Part & Accessories	SK:	R^2	.23	.81	.84	.92	.84	.89
		DW	.81	.71	.93	1.5	.92	1.5
	SL:	R^2	.49	.43	.88	.84	.90	.86
		DW	1.6	1.8	1.5	1.8	1.3	1.7
Manufacture & Assembly of Bicycles	SK:	R^2	.44	.51	.89	.86	.91	.87
		DW	1.7	1.8	1.3	1.7	1.1	1.6
	SL:	R^2	.24	.36	.71	.70	.85	.85
		DW	1.7	1.8	1.6	1.7	1.7	1.8
Manufacture of Professiona & Scientific Equipment	SK:	R^2	.91	.91	.91	.91	.91	.91
		DW	1.7	1.7	1.7	1.9	1.5	2.0
	SL:	R^2	.15	.28	.49	.60	.63	.68
		DW	1.4	1.6	1.4	1.7	1.4	1.6

[a]SK is cost share of capital equation
[b]SL is cost share of labor equation
[c]DW is durbin-Watson statistics

Canning, in Model [B4], $R^2 = 96$; D-W statistic is 1.8. In Model [B6], $R^2 = .96$ and D-W statistics is 1.6. Similarly, for Cement and Concrete Manufacturing, in Model [B4], $R^2 = .96$, D-W Ststistic is 1.1 while in Model [B6], $R^2 = .95$ and D-W Statistic is 1.2.

Theoretically, in Model [B4], the variables are real values deflated by the consumer price index which will eliminate biases due to inflation and cyclical price movements. In Model [B6], on the other hand, the variables are also real values deflated by the industrial production index which take into account biases due to under-utilisation of capacity. In order to make comparisons with the estimates based on the CES production function, empirical results based on Model [B4] are reported. In both cases, the parameter estimates were derived from real values deflated by the consumer price index and corrected for auto-correlation by the AR (1) method. As such we can expect some consistency in the potential biases of the parameter coefficients of the CES production function and the translog cost function. Furthermore, Model [B4] exhibits good statistical performance in terms of the F - statistics, the parameter estimates and the variance covariance matrix. Except for Palm Kernel Oil manufacturing, Leather and Leather Products Industries, the coefficient of determination are high and statistically significant at 99 percent confidence level. The Durbin-Watson statistics for all industries based on Model [B4] are above 1.1 and below 2.3.

Basing on Model [B4], Table 4.2 presents the unrestricted parameter estimates, asymptotic t-ratios and standard errors, and the log of likelihood function of the multivariate translog cost function for the Malaysian manufacturing sector, 1968 - 1984. Most of the coefficients are statistically significant at the 95% percent confidence level and the standard errors are small. The statistical results shown in Table 4:1 and 4:2 indicate a good fit for the systems of equations for the translog cost estimation in the Malaysian manufacturing sector.

Of the 300 estimated parameters, approximately 75 percent are significant at 10% percent or higher. For a structural model of this magnitude, the results are very encouraging.

Slope parameters in the cost share equations reflect changes in the cost shares resulting from changes in logarithmic prices in real terms. The slope coefficients may either be positive or negative, since the second derivatives of lnC with respect to ln P_i and ln P_j may be of either sign. If the cost share is inversely related to ln P_j, i.e. $\beta_{ij} < 0$, this suggests that i and j are substitute inputs. If the cost share increases with a rise in the real price, i.e. $\beta_{ij} > 0$, this suggests that input substitution is limited.

As depicted in Table 4:2, the cost share of the labor parameter estimates (β_{LK}) show that substitution between capital and labor is rather limited in 36 out of the 50 industries studied.

Table 4.2. Unrestricted SUR parameter estimates of capital-labor translog cost function for Malaysian manufacturing sector, 1969-1984

	k[a]	β_{KK}	β_{KL}	L[a]	β_{LK}	β_{LL}
Slaughtering Preparing & Preserving Meat (15)	18.9729** (3.373) (5.625)	2.6646** (4.669) (.571)	−1.1667 −(1.121) (1.040)	3.7333** (3.487) (1.07)	.186412* (1.716) (.108)	.220717 (1.115) (.198)
			Log of likelihood function = −21.0041			
Ice Cream Manufacturing (16)	5.9098** (3.6779) (1.6069)	.51779** (6.4667) (.08007)	−.19835 −(.8219) (.24134)	2.4198** (4.208) (.57506)	−0.148 −(.51571) (.02866)	.2465** (2.8534) (.08637)
			Log of likelihood function = 19.3469			
Manufacture of Other Dairy Products (16)	14.0369** (6.5979) (2.1274)	.6936 (1.1804) (.5876)	.43239 (.5669) (.7626)	2.3765** (5.1509) (.4614)	−.3304** −(2.5928) (.1274)	.7215** (4.3627) (.1654)
			Log of likelihood function = 13.1707			
Pineapple Canning (16)	21.3368** (13.8486) (1.5407)	2.1128** (16.416) (.1287)	−.32822 −(1.230) (.2668)	13.3315** (11.140) (1.1967)	.67331** (6.735) (.0999)	.7044** (3.399) (.207)
			Log of likelihood function = 13.8715			
Other Canning and Preserving of Fruits and Vegetables (16)	31.5036** (6.0836) (5.1785)	.0003 (.09295) (.0003)	.0003 (1.444) (.0002)	8.0204** (7.733) (1.0371)	.0006 (.1080) (.0005)	.0007* (1.849) (.0004)
			Log of likelihood function = −107.287			
Coconut Oil Manufacturing (16)	8.4517** (5.2377) (1.6137)	.9397** (6.6235) (.1419)	−.4453* −(1.8903) (.2356)	2.2246** (8.4264) (.2640)	.01912 (.8239) (.02321)	.2031** (5.2681) (.03856)
			Log of likelihood function = 42.6162			
Palm Oil Manufacturing (12)	13.9544** (12.2761) (1.1367)	.8648** (8.4823) (.1020)	−.0458 −(.3298) (.1388)	.6296** (2.8447) (.2213)	.0206 (1.039) (.0199)	.0104 (.3865) (.0270)
			Log of likelihood function = 8.2686			
Palm Kernel Oil Manufacturing (12)	4.0669** (2.9108) (1.3972)	1.5000** (4.9975) (.3002)	−1.8157** −(3.3040) (.5495)	.7694 (.7232) (.0544)	.394 (.0577) (.0996)	.0576
			Log of likelihood function = 15.8661			
Vegetable and Animal Oils & Fats (16)	2.7146 (1.7320) (1.5674)	2.3164** (4.2346) (.5470)	−3.3437** −(3.3762) (.9904)	3.017** (15.2255) (.1982)	−.1840** −(2.6606) (.06916)	.6129** (4.8753) (.1252)
			Log of likelihood function = 2.4453			
Rice Milling (16)	11.0983** (12.4965) (.8881)	.1328** (8.9178) (.0149)	.9042** (8.9963) (.1005)	3.0661** (14.1080) (.2173)	.0085** (2.3203) (.0036)	.2857** (11.6151) (.0246)
			Log of likelihood function = 135.791			
Biscuit Factories (16)	8.7965** (7.8893) (1.1150)	1.3952** (4.6068) (.3029)	−1.011** −(2.0247) (.4991)	5.7989** (9.1958) (.6306)	.2288 (1.3357) (.1713)	.2834 (1.0038) (.2823)
			Log of likelihood function = 20.7105			

72 Substitutability in Malaysian Manufacturing

Table 4.2. Continued

	k[a]	β_{KK}	β_{KL}	L[a]	β_{LK}	β_{LL}
Sugar Factories and Refineries (14)	18.3074** (10.7877) (1.6971)	1.5844** (9.7310) (.1628)	−.3601** −(2.1762) (.1655)	4.9274** (10.3778) (.4748)	−.2841** −(6.2371) (0.455)	.9036** (19.5181) (.04629)
			Log of likelihood function = 6.31046			
Manufacture of Cocoa, Chocolate and Sugar Confectionery (16)	197.835** (7.5969) (26.0412)	18.4582** (6.5763) (2.8068)	−18.4582** −(6.5763) (2.8068)	40.1026** (6.2094) (6.4.584)	3.5268** (5.0665) (.6961)	−3.5268** −(5.0665) (.6961)
			Log of likelihood function = −76.2793			
Ice Factories (16)	7.2953** (16.3960) (.4449)	1.8268** (3.8299) (.4770)	−1.8089** −(2.9625) (.6106)	4.0558** (15.7192) (.2580)	−.0062 −(.0222) (.2766)	.4045 (1.1426) (.3541)
			Log of likelihood function = 2.03499			
Coffee Factories (15)	7.3639** (6.9108) (1.0656)	1.3944* (4.6211) (.3017)	−1.0970** −(2.2636) (.4846)	2.6167** (7.7499) (.3376)	.2693** (2.8165) (.0956)	−.07990 −(.5203) (.1536)
			Log of likelihood function = 29.7561			
Meehoon, Noodles & Related Products (15)	7.7753** (13.8105) (.5630)	1.0117** (4.1034) (.2465)	−.6090* −(1.9075) (.3193)	5.6405** (13.0846) (.4311)	−.5962** −(3.1582) (.1888)	1.3711** (5.6089) (.2445)
			Log of likelihood function = 3.9187			
Manufacture of Prepared Animal Feeds (16)	5.9534** (10.9926) (.5416)	.3659** (2.3868) (.1533)	.1472 (.0567) (.2596)	1.4889** (11.1455) (.1336)	.0064 (.1699) (.03782)	.1252* (1.9551) (.06404)
			Log of likelihood function = 52.6417			
Soft Drinks and Carbonated Beverages (16)	24.9554** (10.5155) (2.3732)	1.2475** (2.3822) (.5237)	.9997 (1.1355) (.8804)	7.8579** (9.9019) (.7936)	.2283 (1.3039) (.1751)	.5643* (1.9167) (.2944)
			Log of likelihood function = −6.2887			
Tobacco Manufacturing (16)	23.3025** (4.2384) (5.4979)	2.3738** (3.6380) (.6525)	−.9242 −(.8519) (1.0849)	3.30259** (3.7628) (.8777)	.27085** (2.6002) (.1042)	−.0545 −(.3149) (.1732)
			Log of likelihood function = −9.9974			
Manufacture of Leather & Leather Products (15)	14.2353** (7.6682) (1.8564)	1.4988** (7.2727) (.2061)	−.07817 −(.4421) (.1768)	4.1525** (3.6144) (1.1489)	.3364** (2.6376) (.1275)	.04200 (.3839) (.1094)
			Log of likelihood function = −10.8739			
Sawmilling (16)	188.901** (9.1146) (2.7251)	5.5921* (1.9891) (2.8113)	21.5325** (3.0183) (7.1339)	83.5976** (13.2349) (6.3164)	−2.6908** −(3.1404) (85.68)	17.2579** (7.9375) (2.1743)
			Log of likelihood function = −87.3251			
Planning Mills & Joinery Works (16)	11.7243** (13.2667) (.8837)	.3011** (2.7364) (.1831)	.4185* (1.3894) (.3012)	7.9129** (19.0755) (.4148)	−.0616 −(.7163) (.08595)	.9013** (6.3750) (.1414)
			Log of likelihood function = −5.0111			

Table 4.2. Continued

	k^a	β_{KK}	β_{KL}	L^a	β_{LK}	β_{LL}
Manufacture of Furniture & Fixtures (15)	14.1262** (17.7427) (.7962)	1.0653** (7.4380) (.1432)	–.0366 –(.2978) (.1228)	6.1331** (12.8904) (.4758)	.0348 (.4071) (.0856)	.4800** (6.5399) (.0734)
			Log of likelihood function = –10.3312			
Clothing Factories (15)	–5.2379 –(1.6241) (3.2252)	6.5555** (11.6122) (.5645)	–8.7703** –(9.3779) (.9352)	7.8439** (10.6078) (.1294)	–.2481* –(1.9168) (.1294)	.9731** (4.5383) (.2144)
			Log of likelihood function = –10.3312			
Paper & Paper Products (15)	10.3181** (2.6938) (3.8303)	2.3621** (4.7270) (.4997)	–2.1879** –(2.6695) (.8196)	3.6914** (9.3449) (.3950)	.09402* (1.8243) (.0515)	.2036** (2.4083) (.0845)
			Log of likelihood function = –14.4170			
Printing Publishing & Allied Industries (16)	–40.4630** –(4.0362) (10.0250)	11.7977** (13.0632) (.9031)	–19.5471** –(12.1147) (1.6135)	12.3229** (15.5783) (.7910)	.1464** (2.0541) (.07126)	.9695** (7.6149) (.1273)
			Log of likelihood function = –28.5616			
Basic Industrial Chemicals (16)	9.1435** (4.3909) (2.0824)	1.0567** (7.2040) (.1467)	–.6099* –(1.7677) (.3450)	3.5818** (9.6691) (.3704)	–.01166 –(.4467) (.02609)	.3599** (3.8636) (.06138)
			Log of likelihood function = 13.9003			
Chemical Fertiliser & Pesticides (14)	10.8391** (12.5532) (1.0786)	.9220** (7.6995) (.5482)	–.2611** –(1.9187) (.8844)	4.7520** (4.2721) (.2753)	–1.1364** –(7.3666) (.1753)	2.1504** (12.2661)
			Log of likelihood function = –5.6503			
Paints, Varnishes & Lacquers (16)	19.6071** (18.1783) (1.0786)	.7671 (1.3992) (.5482)	1.0949 (1.2381) (.8844)	5.5926** (20.3259) (.2753)	–.1383 –(.9888) (.1399)	.8424** (3.7320) (.2257)
			Log of likelihood function = 22.2485			
Manufacture of Drugs, Medicine & Pharmaceuticals (16)	22.0125** (24.4562) (.9001)	1.8906** (18.6222) (.1015)	–.0004 –(.3395) (.0001)	5.7905** (19.4741) (.2973)	.4402** (12.7833) (.0344)	.0006** (2.6771) (.0002)
			Log of likelihood function = 20.9702			
Soap and Cleaning Preparations (16)	36.0938** (16.7193) (2.1588)	3.3003** (7.6179) (.4332)	–.3065 –(.5972) (.5133)	7.0091** (29.5960) (.2368)	.3570** (7.5123) (.04753)	.3360** (5.9679) (.05631)
			Log of likelihood function = –6.4532			
Perfumes, Cosmetics & Toiletteries (16)	48.1074* (1.9399) (24.7990)	4.6196* (1.8488) (2.4988)	–3.8635 –(.5939) (6.5218)	5.4346** (2.9185) (1.8621)	.4667** (2.4875) (.1876)	–.2023 –(.4130) (.4897)
			Log of likelihood function = –24.2608			
Petroleum Refineries (16)	13.5253** (4.4242) (3.0571)	.9513** (4.3583) (.2183)	–.07394 –(.2276) (.3248)	1.1791** (5.7–38) (.2067)	.02039 (1.3820) (.01476)	.09214* (4.1948) (.02197)
			Log of likelihood function = 15.0864			

74 Substitutability in Malaysian Manufacturing

Table 4.2. Continued

	k^a	β_{KK}	β_{KL}	L^a	β_{LK}	β_{LL}
Petroleum & Coal Products	24.5975	26.8219**	−35.9719**	4.3453*	3.1673**	−4.0684
(14)	(1.3262)	(10.8457)	(−5.7354)	(1.7597)	(9.6199)	−(4.8723)
	(18.5481)	(2.4730)	(6.2719)	(2.4694)	(.3292)	(.8350)
		Log of likelihoof function = −50.6679				
Rubber Products	38.4408**	.00086	.00029	12.3814**	.00013	.00011
(15)	(5.5726)	(.2356)	(1.1169)	(5.3913)	(.1083)	(1.2440)
	(6.8982)	(.0003)	(.0003)	(2.2966)	(.0001)	(.0009)
		Log of likelihood function = −107.761				
Plastic Products	11.9493**	1.4897**	−.8795	6.2165**	.0058	.5478**
(14)	(7.0256)	(4.4713)	−(1.5918)	(7.7755)	(.03693)	(2.1093)
	(1.7008)	(.3332)	(.5525)	(.7995)	(.1566)	(.2597)
		Log of likelihood function = 10.7986				
Pottery China and	5.6948	8.4198**	−10.8266	3.4271	3.4504**	−4.2026**
Earthernware	(1.2273)	(6.5195)	−(5.6210)	(1.4708)	(6.0824)	−(5.0254)
(16)	(4.6403)	(1.2915)	(1.9261)	(.3306)	(.5673)	(.8363)
		Log of likelihood function = 37.2425				
Hydraulic Cement	23.8986**	2.1158**	−.4922	2.6256**	.1871**	−.0088
(16)	(10.7872)	(7.3099)	−(1.1066)	(19.3976)	(10.5819)	−(.3237)
	(2.2155)	(.2894)	(.4448)	(.1354)	(.0176)	(.0272)
		Log of likelihood function = 6.0848				
Cement & Concrete	15.2463**	2.7781**	−2.2171**	5.4039**	.1588*	.3176**
(16)	(12.0617)	(9.8954)	−(5.4484)	(13.3905)	(1.7712)	(2.4446)
	(1.2540)	(.2807)	(.4069)	(.4036)	(.08963)	(.1299)
		Log of likelihood function = −4.4913				
Primary Iron & Steels	11.1536**	1.5050**	−1.0696**	4.9827**	.0352	.4701**
Industries	(8.9390)	(13.5826)	−(4.1199)	(10.9971)	(.8742)	(4.9864)
(15)	(1.2477)	(.1108)	(.2596)	(.4531)	(.0402)	(.0943)
		Log of likelihood function = 18.2763				
Non–Ferrous Metal	10.5066**	2.6251**	−2.7021**	3.3309**	.06365	.2361
Products	(12.9604)	(7.4587)	−(5.1889)	(12.7597)	(.5616)	(1.4078)
(16)	(.8107)	(.3519)	(.5207)	(.2610)	(.1133)	(.1677)
		Log of likelihood function = −3.3469				
Wire Products	16.3093**	−1.0004**	1.6596**	−3.9449**	−.0001**	.4007**
(16)	(6.6775)	−(2.7204)	(5.4504)	(11.3723)	−(4.6173)	(9.2661)
	(2.4424)	(.0002)	(.3045)	(.3469)	(.0002)	(.0432)
		Log of likelihood function = −7.5823				
Brass, Copper Pewter &	14.6139**	1.4793**	−.5015	6.6736**	.1653	−.4752**
Alluminium Product	(11.2359)	(4.5209)	−(.8979)	(14.3693)	(1.4146)	(2.3827)
(16)	(1.3006)	(.3272)	(.5586)	(.4644)	(.1168)	(.1995)
		Log of likelihood function = 4.7860				
Industrial Machinery &	25.7625**	.10007	2.9697**	13.2794**	−.9336**	2.8998**
Parts	(11.5296)	(.1729)	(3.3621)	(17.9883)	−(4.8815)	(9.9369)
(16)	(2.2345)	(.5789)	(.8833)	(.7382)	(.1913)	(.2918)
		Log of likelihood function = 17.2311				

Table 4.2. Continued

	k[a]	β_{KK}	β_{KL}	L[a]	β_{LK}	β_{LL}
Electrical Machinery,	13.6881**	1.6805**	−1.0928**	3.1783**	.1973**	−.0391
Apparatus & Appliances	(11.2817)	(4.5920)	−(2.4624)	(15.8560)	(3.2638)	−(.5332)
(15)	(1.2133)	(.3660)	(.4438)	(.2004)	(.0605)	(.0733)
		Log of likelihood function = −.7509				
Shipbuilding Boatmaking	9.6464**	2.4901**	−2.7078**	7.9286**	.2469	.5107**
& Repairing	(7.4499)	(6.6583)	−(4.7386)	(15.3323)	(1.6530)	(2.2377)
(16)	(1.2948)	(.3740)	(.5714)	(.1494)	(.1494)	(.2282)
		Log of likelihood function = −28.3847				
Manufacture of Motor	19.5915**	1.7289**	−1.3205	3.5089	.2195	.7111
Vehicle Bodies	(5.4778)	(4.5344)	−(2.3007)	(.9098)	(.5340)	(1.1489)
(16)	(3.5765)	(.3812)	(.5739)	(3.8566)	(.4110)	(.6189)
		Log of likelihood function = 1.0970				
Manufacture of Motor	7.3433**	1.5685**	−1.4635**	4.0248**	.1606	.1818
Vehicle Part & Accessories	(16.3399)	(3.3462)	−(2.4460)	(15.5037)	(.7188)	(.6525)
(16)	(.4494)	(.4687)	(.5986)	(.2596)	(.2234)	(.2786)
		Log of likelihood function = 83.6252				
Manufacture & Assembly	15.1477**	1.3309**	−.2688	1.5123	.04614	.3657
of Bicycles	(3.8603)	(3.2392)	−(.4103)	(1.1052)	(.3221)	(1.6007)
(16)	(3.9240)	(.4109)	(.6552)	(1.3683)	(.1433)	(.2285)
		Log of likelihood function = 10.9122				
Manufacture of Professional	15.8127**	2.6955**	−1.9511**	4.0495**	.4590*	−.2989
and Scientific Equipment	(12.6179)	(6.7031)	−(3.3129)	(5.3361)	(1.8848)	−(.8382)
(16)	(1.2532)	(.4021)	(.5889)	(.7589)	(.2435)	(.3566)
		Log of likelihood function = −22.9675				

[a]For each industry group, row one shows the coefficients, row two shows the t-statistics and row three shows the standard errors of each coefficient.

**implies significant at .05 with a 2-tailed t-test

* implies significant at 10 with a 2-tailed t-test

When restrictions of linear homogeneity and symmetry constraints are imposed, one equation becomes redundant and only one equation is estimated. Based on Model [B4], the equation with superior statistical results in terms of R^2, D-W statistics, variance - covariance matrix, F - statistics and the β - coefficients were chosen for each industry. In Ice cream Manufacturing for example, the cost share of capital was chosen instead of cost share of labor equation. On the other hand, in the case of Manufacture of Other Dairy Products, the cost share of labor equation was chosen. Elasticity of substitution estimates are based on the chosen equation under restriction.

Several tests are also conducted to determine if the model [B4] estimated is compatible with the neoclassical theory of cost and production. Further restrictions implied by economic theory include the monotonicity and concavity of the cost function. Neither monotonicity nor concavity of the cost function with respect to input prices will be satisfied globally, however, since the translog specification is only an approximation to the true cost function. Sufficient conditions for these are positive fitted cost shares and negative definiteness of the second order partials of the cost share equations. The fitted cost shares from OLS or SUR estimates are positive for all industries. This can be verified from Table 4:2. However, not all diagonal elements of a Hessian Matrix, i.e. β_{KK} and β_{LL} are negative. This implies that in a number of industries, the cost function is not concave in input prices.

Our primary objective in this section is to measure the elasticities of substitution between capital and labor in 5-digit Malaysian manufacturing industries. Basing on Model [B4], the Allen partial elasticity of substitution between capital and labor was calculated at the mean cost shares, at begining year of analysis (1969), base year (1980) and at ending year of analysis (1984). The elasticity of substitution are reported in Table 4:4.

Interestingly, the elasticities evaluated at the mean cost shares and in different years show insignificant variations. In 16 cases, the elasticity increases very slightly between 1968 and 1984. Evaluated at mean cost shares, more than half of the elasticities are numerically less than unity. In 32 percent of the cases, the value of the elasticity is less than 0.8. In 14 percent of the cases, it exceeds 1.10 while in more than 50 percent of the cases, it lies between 0.8 and 1.10. The elasticity of substitution between labor and capital in 5-digit Malaysian manufacturing industries ranged between -3.716 for clothing industries to 4.649 for petroleum refining. This is a most interesting result since they confirm similar findings based on the CES production function estimation. In the earlier estimation of the CES production function, the range was -1.27 and 3.51

Table 4.3. Restricted estimates of capital-labor translog cost function for Malaysian manufacturing sector, 1969-1984

	k[a]	β_{KK}	β_{KL}	L[a]	β_{LK}	β_{LL}
Slaughtering, Preparing & Perserving Meat	−2.7333	1.1667	.1864	3.7333	.1864	.2207
Ice Cream Manufacturing	5.9098	.5178	−.1984	−4.9098	−.1984	.0148
Manufacture of Other Dairy Products	−1.3765	−.4324	−.3304	2.3765	−.3304	.7215
Pineapple Canning	−12.3315	.3282	.6733	13.3315	.6733	.7044
Other Canning and Perserving of Fruits & Vegetables	−7.0204	−.0003	.0006	8.0204	.0006	.0007
Coconut Oil Manufacturing	8.4517	.9397	−.4453	−7.4517	−.4453	−.01912
Palm Oil Manufacturing	13.9544	.8648	−.0458	−12.9544	−.0458	−.0206
Palm Kernel Oil Manufacturing	4.0669	1.5000	−1.8157	−3.0669	−1.8157	−.0394
Vegetable and Animal Oils & Fats	−2.0170	3.3437	−.1840	3.017	−.1840	.6129
Rice Milling	−2.006	−.9042	.0085	3.0661	.0085	.2857
Biscuit Factories	8.7965	1.3952	−1.001	−7.7965	−1.011	−.2288
Sugar Factories & Refineries	−3.9274	.3601	−.2841	4.9274	−.2841	.9036
Manufacture of Cocoa, Chocolate & Confectionery	−39.1026	18.4582	3.5268	40.1026	3.5268	−3.5268
Ice Factories	7.2953	1.8268	−1.8089	−6.2963	−1.8089	.0062
Coffee Factories	7.3639	1.3944	−1.0970	−6.3639	−1.0970	−.2693
Meehoon, Noodles & Related Products	−4.6405	.6090	−.5962	5.6405	−.5962	1.3711
Manufacture of Prepared Animal Feeds	−0.4889	−.0147	.0064	1.4889	.0064	.1252
Soft Drinks & Carbonated Beverages	−6.8579	−.9997	.2283	7.8579	.2283	.5643
Tobacco Manufacturing	−2.3026	.9242	.2709	3.3026	.2709	−.0545
Manufacture of Leather & Leather Products	−3.1525	.0782	.3364	4.1525	.3364	.0420
Sawmilling	−85.7503	−21.5325	−2.6908	83.5976	−2.6908	17.2579
Planing Mills & Joinery Works	−6.9129	−.4185	−.0616	7.9129	−.0616	.9013
Manufacture of Furniture & Fixtures	−5.1331	.0366	.0348	6.1331	.0348	.4800
Clothing Factories	−6.8439	8.7703	−.2481	7.8439	−.2481	.9731
Paper & Paper Products	10.3181	2.3621	−2.1819	−9.3181	−2.1879	−.0940
Printing Publishing & Allied Industries	11.3229	19.5471	.1464	12.3229	.1464	.9695
Basic Industrial Chemicals	−2.5818	.6099	−.0116	3.5818	−.0116	.3599
Chemical Fertiliser & Pesticides	−3.7520	.2611	−1.1364	4.7520	−1.1364	2.1504
Paints Varnishes & Lacquers	−4.5926	−1.0949	−.1383	5.5926	−.1383	.8424

Table 4.3. Continued

	k^a	β_{KK}	β_{KL}	L^a	β_{LK}	β_{LL}
Manufacture of Drugs, Medicine & Pharmacueticals	−4.7905	.0004	.4402	5.7905	.4402	.0006
Soaps & Cleaning Preparations	−6.0091	.3065	.3570	7.0091	.3570	.3570
Perfumes, Cosmetics & Toiletteries	−4.4346	3.8635	.4667	5.4346	.4667	−.2023
Petroleum Refineries	−0.1791	.6739	.0204	1.1791	.0204	.0924
Petroleum & Coal Products	−3.3453	35.9719	3.1673	4.3453	3.1673	−4.0684
Rubber Products	−11.3814	−.0003	.0001	12.3814	.0001	.0001
Plastic Products	−5.2165	.8795	.0058	6.2165	.0058	.5478
Pottery, China & Earthernware	−2.4271	10.8266	3.4504	4.4271	3.4504	−4.2026
Hydraulic Cement	−1.6256	.4922	.1871	2.6256	.1871	−.0088
Cement & Concrete	15.2463	2.7781	−2.2171	−14.2463	−2.2171	−.1588
Primary Iron & Steel Industries	11.1536	1.5050	−1.0696	−10.1536	−1.0696	−.0352
Non Ferrous Metal Products	10.5066	2.6251	−2.7021	−9.5066	−2.7021	−.0636
Wire Products	−2.9449	.0004	−.0001	3.9449	−.0001	.4007
Brass Copper Pewter & Alluminium Products	−5.6736	.5015	.1653	6.6736	.1653	.4752
Industrial Machinery & Parts	−12.2794	−2.9697	−.9336	13.2794	−.9336	2.8998
Electrical Machinery, Apparatus & Appliances	13.6881	1.6805	−1.0928	−12.6881	−1.0928	−.1973
Shipbuilding, Boatmaking & Repair	9.6464	2.4901	−2.7078	−8.6464	−2.7078	−.2469
Manufacture of Motor Vehicle Bodies	19.5915	1.7289	−1.3205	−18.5915	−1.3205	−.2195
Manufacture of Motor Vehicle Parts & Accessories	7.3433	1.5685	−1.4635	−6.3433	−1.4635	−.1606
Manufacture of Assembly of Bicycles	15.1477	1.3309	−.2688	−14.1477	−.2688	−.0461
Manufacture of Professional & Scientific Equipment	15.8127	2.6955	−1.9511	−14.8127	−1.9511	−.4590

Table 4.4. Elasticity of substitution in Malaysian manufacturing industries: a translog cost function approach

	TLC (M)[a]	TLC (1968)[b]	TLC (1980)[c]	TLC (1984)[d]
Slaughtering, Preparing and Preserving Meat	−3.54	0.925	0.615	0.645
Ice Cream Manufacturing	0.965	0.995	0.923	0.905
Manufacture of Other Dairy Products	0.812	0.866	0.464	0.395
Pineapple Canning	1.335	1.064	1.309	1.406
Other Canning and Perserving of Fruits & Vegetables	1.00	1.00	1.00	1.00
Coconut Oil Manufacturing	2.922	1.055	1.128	1.328
Palm Oil Manufacturing	0.950	0.968	0.852	0.582
Palm Kernel Oil Manufacturing	0.679	0.501	−1.941	−1.478
Vegetable and Animal Oils & Fats	0.772	0.961	0.525	0.446
Rice Milling	1.019	1.018	1.027	1.044
Biscuit Factories	0.667	0.869	0.537	0.20
Sugar Factories & Refineries	0.664	0.910	0.256	0.335
Manufacture of Cocoa, Chocolate & Confectionery	0.970	0.987	0.943	0.884
Ice Factories	0.938	1.00	0.996	0.995
Coffee Factories	0.223	0.614	0.268	0.243
Meehoon, Noodles & Related Products	0.371	0.936	0.572	0.545
Manufacture of Prepared Animal Feeds	1.089	1.006	1.017	1.030
Soft Drinks & Carbonated Beverages	1.339	1.012	1.073	1.087
Tobacco Manufacturing	1.325	1.051	1.136	1.271
Manufacture of Leather & Leather Products	0.980	0.986	0.987	0.937
Sawmilling	0.955	0.849	0.180	0.134
Planning Mills & Joinery Works	0.952	0.992	0.976	0.964
Manufacture of Furniture & Fixtures	1.014	1.002	1.007	1.074
Clothing Factories	−1.716	0.660	0.843	−1.038
Paper & Paper Products	1.015	1.010	1.038	1.076
Printing Publishing & Allied Industries	0.230	0.180	−2.044	−2.284
Manufacture of Basic Industrial Chemicals	0.853	0.936	0.788	0.782

Table 4.4. Elasticity of substitution in Malaysian manufacturing industries: a translog cost function approach

	TLC (M)[a]	TLC (1968)[b]	TLC (1980)[c]	TLC (1984)[d]
Manufacture of Chemical Fertiliser & Pesticides	0.110	1.00	0.679	0.128
Manufacture of Paint Varnishes & Lacquers	0.943	0.988	0.953	0.937
Manufacture of Drugs, Medicine & Pharmacueticals	1.041	1.016	1.055	1.55
Manufacture of Soaps & Cleaning Preparations	1.157	1.016	1.073	1.802
Manufacture of Perfumes, Cosmetics & Toiletteries	1.324	0.839	0.170	0.012
Petroleum Refineries	4.649	1.022	1.057	1.263
Petroleum & Coal Products	1.051	1.00	1.131	1.485
Rubber Products	1.00	1.00	1.00	1.00
Plastic Products	0.321	0.933	0.636	0.441
Pottery, China & Earthernware	−0.396	−0.360	−1.896	−3.29
Hydraulic Cement	1.535	1.026	1.164	1.251
Cement & Concrete	1.051	1.137	1.070	1.106
Primary Iron & Steel Industries	0.083	0.550	0.468	1.241
Non Ferrous Metal Products	1.013	1.007	1.040	1.043
Wire Products Manufacturing Brass Copper Pewter &	1.00	1.00	1.00	1.00
Alluminium Products	0.889	0.965	0.845	0.767
Industrial Machinery & Parts Electrical Machinery,	0.462	0.975	0.697	0.287
Apparatus & Appliances	0.67	0.914	0.575	0.272
Shipbuilding, Boatmaking & Repairing	1.039	1.009	1.031	1.054
Manufacture of Motor Vehicle Bodies	0.017	0.959	0.749	0.585
Manufacture of Motor Vehicle Parts & Accessories	0.681	0.828	0.320	0.435
Manufacture of Assembly of Bicycles	1.026	1.006	1.006	1.05
Manufacture of Professional & Scientific Equipment	1.085	1.019	1.132	1.098

[a] Elasticity of substitution calculated at mean real cost shares.
[b] Elasticity of substitution calculated at beginning year of analysis.
[c] Elasticity of substitution calculated at base year=1980.
[d] Elasticity of substitution calculated at ending year of analysis.

(for clothing and petroleum refining, respectively). Another interesting point to be noted is the generally low substitution possibilities between capital and labor inputs measured either by the CES production function approach or the translog cost function approach.

Elasticity of substitution with technical change.

Alternative versions of Model [B7] to Model [B12] are estimated and a statistical search procedure is carried out to determine the best fitting model in order to estimate the effect of technical change. Although Model [B8] shows superior results in terms of R^2 and D-W statistics, the results cannot be reported. None of the coefficients of ln S_T in the 50 industries estimated are statistically significant. Experiments with alternative versions of estimation fail to produce any meaningful results. For example, in Rice Milling and Leather Products Industries, the coefficients are meaningless with small t-ratios and very large standard errors. Since respecification of this model also implies respecification of the complete system of equations, the estimates with technical change are not reported.

	a_T	KT	LT	TT
Rice Milling	-.371840	6.27418	-.317636	.505043
	(-4.13876)	(.61952)	(-4.12965)	(.865292)
	(358.7)	(1231.6)	(203.7)	(1032.6)
Leather & Leather Products	-95103.7	-4037.02	-6934.01	-168.926
	(8.93646)	-(6.81428)	-(6.89419)	-(1.5429)
	(10642.2)	(592.43)	(1005.78)	(109.486)

The theoretical and empirical model of the CES translog cost function

The third specification is the CES - translog cost function, which was recently developed by Pollack, Sickles and Wales (1984). The CES - translog cost function combines the CES production function and the translog cost function. Like the translog, it is a flexible form, but it is compatible with a wider range of substitution possibilities compared to either the CES or the translog cost function. Since the CES - translog includes both the CES production function and the translog cost function

82 Substitutability in Malaysian Manufacturing

as special cases, it permits nested testing using conventional statistical techniques.

The CES - translog cost function pivots off the CES production function and the translog cost function by adding another parameter to the translog cost function. Cost minimisation entails

$$l = -P_L L - P_K K + \lambda \{[\delta K^{-\rho} + (1-\delta)L^{-\rho}]^{-v/\rho} - Y^*\}$$

and the first order conditions,

$$\frac{\delta l}{\delta K} = \delta K^{-(\rho+1)} Y^{-(v/\rho+1)} - P_K = 0$$

$$\frac{\delta l}{\delta L} = (1-\delta)L^{-(\rho+1)} Y^{-(v/\rho+1)} - P_L = 0$$

$$\frac{\delta l}{\delta \lambda} = [\delta K^{-\rho} + (1-\delta)L^{-\rho}]^{-v/\rho} - Y^* = 0$$

$$MRST_{K,L} = \frac{\delta K^{-(\rho+1)}}{(1-\delta)L^{-(\rho+1)}} = \frac{P_K}{P_L}$$

$$\frac{K}{L} = [(\frac{1-\delta}{\delta})\frac{P_L}{P_K}]^{1/1+\rho} \text{ where } \sigma = \frac{1}{1+\rho}$$

$$= [(\frac{1-\delta}{\delta})\frac{P_L}{P_K}]^{\sigma}$$

Solving simultaneously the first order conditions, gives optimal factor demand functions,

$$L^*(P_K, P_L, Y) = Y^{1/v} [\delta + (1-\delta)(\frac{K}{L})^{-\sigma}]^{1/\sigma}$$

$$K^*(P_K, P_L, Y) = Y^{1/v} [(1-\delta) + \delta(\frac{L}{K})^{-\sigma}]^{1/\sigma}$$

$$C^*(P_K, P_L, Y) = P_K K^* + P_L L^k$$

$$= P_K Y^{1/\nu}[(1-\sigma) + \delta(\frac{L}{K})^{-\sigma}]^{1/\sigma} + P_L Y^{1/\nu}[\delta + (1-\delta)(\frac{K}{L})^{-\sigma}]^{1/\sigma}$$

$$= P_L L^* [1 + (\frac{P_K}{P_L})(\frac{1-\delta}{\delta})(\frac{P_L}{P_K})]^\sigma \qquad (4:26)$$

Substituting L^* and $\rho = \dfrac{1-\sigma}{\sigma}$, we have the sepcific CES Cost function of the form;

$$C^*(P_K, P_L, Y) = \delta^{1/\rho} Y^{1/\nu}[P_L^{1-\sigma} + P_K^{1-\sigma}(\frac{1-\delta}{\delta})^\sigma]^{1/1-\sigma}$$

The first order approximation of the CES - translog cost function takes the form,

$$\ln C(P_K, P_L, Y) = 1/\rho \ln \delta + 1/\nu \ln Y$$

$$+ \ln [P_L^{1-\sigma} + P_K^{1-\sigma}(\frac{1-\delta}{\delta})]^{1/1-\sigma}$$

$$+ 1/2 \sum_i \sum_j \beta_{ij} \ln P_L \ln P_K$$

$$+ \sum_i \beta_{iY} \ln Y \ln P_L \ln P_K \qquad (4:27)$$

Assuming homotheticity, the CES-translog cost function is of the form,

$$\ln C(P_K, P_L) = a_o + \ln [\sum_i a_i P_i^{1-\sigma}]^{1/1-\sigma}$$

$$+ 1/2 \sum_i \sum_j \beta_{ij} \ln P_K \ln P_L \qquad (4:28)$$

where $a_o = 1/\rho \ln \delta$

$$\sum_i a_i = 1$$

$$\beta_{ij} = \beta_{ji}$$

$$\sum_i \beta_{ij} = 0$$

Invoking Shephard's Lemma,

$$\frac{\delta C(P_i, P_j)}{\delta P_i} = X_i^*, \text{ the cost - minimizing input demand}$$

and $\dfrac{\delta \ln C(P_i, P_j)}{\delta \ln P_i} = \dfrac{P_i X_i}{C} = S_i$

Thus, in terms of cost shares (S_i), the factor demand functions are of the form,

$$S_i = \left[\frac{a_i P_i^{1-\sigma}}{a_i P_j^{1-\sigma}}\right] + \sum_j \beta_{ij} \ln P_j \qquad (4{:}29)$$

The Cross Allen Elasticities (AES) are now given by

$$\sigma_{ij} = \{ [(\sigma^* - 1) \left[\frac{a_i P_i^{1-\sigma}}{a_j P_j^{1-\sigma}}\right] \times \left[\frac{a_j P_j^{1-\sigma}}{a_i P_i^{1-\sigma}}\right]$$

$$+ \beta_{ij} + S_i S_j \} / S_i S_j$$

where $i \# j$

and $i, j = K, L$

σ^* is the elasticity of substitution calculated from the CES production function.

For the two-factor case, the estimation equations for the CES-translog cost function are as follows,

$$S_K = a_K + \beta_{KK} \ln P_K + \beta_{KL} \ln P_L + \epsilon_k \qquad (4:30)$$

$$S_L = a_L + \beta_{LK} \ln P_K + \beta_{LL} \ln P_L + \epsilon_L \qquad (4:31)$$

where $a_K = \left[\dfrac{a_L P_L^{1\sigma}}{a_K P_K^{1\sigma}}\right]$

$$a_L = \left[\dfrac{a_K P_K^{1-\sigma}}{a_L P_L^{1-\sigma}}\right]$$

and

$$\beta_{KL} = \{[(\sigma^*-1) a_K \cdot a_L] + \beta_{KL} + S_K S_L\}/S_K S_L$$

Note that equations [4:30] [4:31] are exactly the same as equations [4:16] [4:17]. Thus the same estimation procedures as the translog cost function can be used to recalculate the elasticity of substitution derived from the CES - translog cost function.

Attempts were made to recalculate the elasticity of substitution based on the CES - translog cost function at mean cost shares for the beginning and ending years of analysis as well as at base year, 1980. The results of both attempts at calculation, based on unrestricted as well as restricted parameter estimates, could not be validated. The elasticities based on CES - translog cost function approach seem to differ widely compared to the elasticities based on CES production function and the translog cost function approaches. For example, in the Manufacture of Perfumes, Cosmetics and Toiletteries, the CES σ = 0.59 TLC σ = 1.32 while the CES - TLC σ = -29.7. Similarly, in the case of Manufacture of Drugs, Medicines and Pharmacueticals, CES σ = 0.06, TLC σ = 1.04 while the CES - TLC σ = -50.8. The only consistent pattern shown by all three alternative measures is the generally low substitutability between capital and labor in the Malaysian manufacturing sector. This conclusion of generally low substitutability is based on the fact that the actual elasticities are probably lower than the calculated elasticities. This is in consideration of numerous biases affecting their estimation as discussed earlier.

CHAPTER V.
CHOICE OF APPROPRIATE TECHNIQUE AND EMPLOYMENT GENERATION

The choice of appropriate techniques presupposes a certain degree, however small, of technological feasibility. If industries are flexible in the sense that numerous techniques with widely diverging factor proportions exist, then policies to reduce the wage-rental ratios, such as wage subsidies or capital - use taxes, can be effective in stimulating firms to choose more labor-intensive techniques. If, on the contrary, in an industry only one technique exists, factor price policies have no effect on the choice of technique. The success of such policies depends on the following conditions,
1. whether technical substitution possibilities exist
2. the effectiveness of policies in changing factor prices
3. the extent to which firms are influenced by price signals in their choice of techniques
4. the availability of factors which influence the choice of appropriate techniques

The basic question which this section tries to answer is whether the economic policies of the Malaysian government and the decisions taken by private local and foreign firms are likely to have influenced the choice of inappropriate techniques in the manufacturing sector.

One of the procedures consists of testing whether capital-intensive industries have been systematically promoted by fiscal incentives and by trade policies. A second method would be that of direct questioning of private firm managers and government planners and executives. Unfortunately, such an undertaking is beyond the scope of this study. This chapter will focus on a discussion based on past literature to shed some understanding on these issues. The first part of the chapter discusses the factors which may influence the choice of technique in a developing country. The second part discusses other possible reasons for the choice of inappropriate techniques and low labor absorption in the manufacturing sector of developing countries such as Malaysia.

Assessment of Technical Substitution Possibilities

The foregoing chapters III and IV have focussed on the measurement of the elasticity of substitution between capital and labor in 5-digit Malaysian manufacturing industries. Chapter III focusses on the CES production function approach while chapter IV focusses on the translog cost function approach. The estimated values of the elasticity of substitution are presented in Tables 3.2 and 4.4, respectively. Most of the industries exhibit elasticities which are significantly different from zero, thus discrediting the notion of fixed-proportions.

The elasticity of substitution is chosen as an indicator of substitution possibilities because it directly answers the question policy makers are interested in, how much can factor proportions be expected to change when factor prices are changed by a given proportion?

Despite differences in the two estimation procedures, the alternative estimates do not produce substantially different results. Although in a number of cases, the two estimates contradict each other with regard to whether the elasticity is greater or lesser than unity, in most industries, the elasticity is rather low. The impression that low substitution elasticities are more common in Malaysia's manufacturing industries is reinforced by Hoffman & Tan's study. Furthermore, due to the upward biases of elasticity measurement as discussed in chapter III, the "actual" elasticities are probably in fact lower than the calculated elasticities.

The practical significance of these results is however not immediately clear due to the underlying assumptions of the production function. One of the major criticims of econometric estimates of elasticity of substitution is the assumption of homogeneous inputs and homogeneous single output [Morawetz, 1974; Gaude, 1975]. When more than one homogeneous product is included in the industry definition, the elasticity of substitution has a different meaning. Substitution is no longer only a matter of choice of technique but also a matter of choice of product-mix. It follows that choice between different factor combinations depends not only on the production function and relative factor prices, but also on consumer demand. Not only the firms, but also the consumers decide how output-mix changes due to changes in factor prices.

In practice however, the separation of the choice of technique from the choice of output involves several problems. First, when appropriate technology is defined in terms of social optimum, it may be necessary to use product-mix as a policy variable in order to achieve employment and distributional goals [Morley & Smith, 1977]. In this sense, it would seem unnecessary to separate the choice of technique from the choice of output-mix. The main argument for separating the choice of technique from the choice of product is to ensure that the effects of factor prices are predictable. If changes in product-mix are also desirable, they may require different or additional policy measures.

In addition to encouraging more appropriate or labor-intensive industries through factor-price manipulation, it may also be possible to encourage the production of more appropriate goods within each industry (rattan rather than glass and steel furniture, as an example). According to Lancaster (1966) and Stewart (1972, 1973), products may be classified in at least three different ways. First, they may be grouped according to cross-elasticities of demand, identical products having infinite cross elasticities. A more practical way however is to group products according to their physical attributes. Thirdly, products may be classified on the basis of needs that they fulfil. The third classification may enable the development of products which are more appropriate to the factor endowments of labor-abundant countries. For example, production of detergents would be more capital-intensive compared to the manufacture of other types of soap.

Despite the conceptual problems mentioned, the range of the elasticity of substitution measured through the CES production function and the translog cost function approaches seem to concur with similar studies in other LDCs. Various factors are identified which may influence the range and density of substitution between capital and labor in manufacturing industries. These factors are discussed below.

Degree of product differentiation.

In most developing countries such as Malaysia with a limited and less diversified domestic market, product differentiation is rather limited. Product differentiation is understood here in a physical sense which has implications for production techniques and factor use. The following example of the tobacco industry is frequently mentioned as technologically flexible (Baranson, 1979). This is so when it includes cigarettes, cigars and other products. While cigarette production is fairly capital-intensive and offers relatively little flexibility in factor proportions, cigar production is more labor-intensive. If the industry includes both products, the range of substitution in the tobacco industry is likely to be large. However, since the Malaysian market is less diversified, the industry includes mainly cigarettes. Although there is quite substantial differentiation in terms of brand names, they essentially involve the same technique of production and factor use.

Depth of transformation.

A second determinant of substitution possibilities is the depth of industrial transformation which can be defined by the number of stages or sub-processes. If there is a single stage with three alternative processes, then there are only three alternative techniques to choose from. If however there are three stages in manufacturing a product, and at each stage three process alternatives exist, then the number of theoretically possible combinations is twenty-seven. In practice the number of transformations will be much smaller for two reasons. First, some alternative combinations may have the same factor proportions. Second, not all combinations are feasible, because they may not fit in terms of quality standards, rates of output, and other dimensions. Some combinations may also be inefficient. In Malaysia, the concern with quality and standards would mean choosing modern techniques, thus reducing further alternative techniques available.

Ease of mechanisation.

In industries with techniques which are easily mechanically operated and controlled, such as chemical industries, technical flexibility is small. Mechanical procedures in such industries are more efficient and cost effective. Highly mechanised processes are also likely to be more integrated than less mechanised one. Consequently, process stage are less distinguishable. Great ease of mechanisation therefore, seems to be associated with low substitutability.

Skill constraint.

The scarcity of skills in LDC manufacturing industries strongly influence the range of factor substitution. The pool of experienced managers, supervisors and technicians are normally supplied by expatriates. Even if they are local personnel, they tend to command relatively higher wages. The choice of techniques is then affected through the high cost of this factor. The lack of skilled technicians and artisans has more constraining consequences. It explains low labor productivity and also explains why certain advanced techniques are inefficient. The lack of specialised and skilled workers tend to reduce the range of substitution in the sense that both most advanced and most labor-intensive techniques may be excluded from the set of efficient techniques.

Factors Responsible for The Choice of Inappropriate Techniques

In conventional economic theory, labor-abundant countries such as Malaysia would use more labor-intensive methods of production while capital-rich countries such as the advanced industrialised countries would use more capital-intensive methods. In practise however, many industries in labor-abundant countries in the developing world including Malaysia tend to be capital-intensive [David Lim, 1975; Maisom Abdullah, 1979].

The empirical investigation on the elasticity of substitution also suggests that inappropriate choice of techniques may have resulted in low substitutability.

A substantial literature has developed recently dealing with the causes of biases in the choice of techniques. It is possible to distinguish two dominant themes in the literature. One attributes non-optimal choices of technique to biases in government policies resulting in price distortions and the other to the behavior of private entreprenuers and in particular that of multinational firms.

The influence of government policy biases

The basic question which needs to be answered is whether economic policies of the Malaysian government are likely to have influenced the choice of inappropriate technologies and in what manner. The large number of possibilities for the government to influence the choice of appropriate technologies can be subdivided by distinguishing three types of intervention.
a) direct intervention through the governments' role as entreprenuer and through public spending,
b) direct intervention through legislation or equivalent policy instruments,
c) indirect intervention through the alteration of market signals.

In general, all interventions have some bearing on prices. However, the biases resulting from the last type of interventions are mainly discussed in the literature as the major reasons for the distortionary effects on factor prices in developing countries [Siggel, 1986]. The price-incentive school stresses the price distortions as a major cause of many of the problems of LDCs, including inappropriate choice of techniques and the consequent unemployment problems. Of major concern are with relatively high wages in the modern sector, relatively low prices of capital, caused by low interest rates, tax incentives related to investment promotion and over-valuation of the exchange rate as well as high levels of protection.

In Malaysia, a wide variety of government policies have made capital artificially cheap. Capital is made cheaper than its true value through government subsidized loans, especially for small and medium-scale industries, and over-valued exchange rates. According to the strict Purchasing Power Parity criteria, the Malaysian Ringgit effective exchange rate was over-valued by 5 percent in 1982, 9 percent in 1984 and 7 percent in 1985 [Gan, 1987]. The price of capital is also affected by low tariffs for imported capital goods, tax holidays and other investment incentives, accelerated depreciation allowances on capital goods and investment tax credits offered by the government to investing firms in

Malaysia. Coupled with relatively high urban wages, the low price of capital has resulted in a higher wage-rental ratio in the manufacturing sector than it's true value in terms of the available factors of production in the country.

Biases in private decision making

The second large group of factors responsible for the choice of inappropriate techniques concerns the behavior of private decision makers as opposed to government policies. Several researchers concerned with the choice of technique in developing countries have argued recently that entreprenuerial or business firm's decisions may be responsible for the use of inappropriate techniques even if the price signals are correct, and even more so under distortionary price signals [Wells, 1973; Lecraw, 1984; Morley & Smith, 1977]. Paul Chan (1979) found that foreign firms in Malaysia tend to obtain their machinery and transfer their technology more or less intact from their parent companies in their home countries. Little adaptation is made to take advantage of the relatively abundant labor in Malaysia. Adaptations made are mainly to suit the requirements of the smaller domestic market. Chan's conclusions concur with findings in Brazil, Pakistan and Peurto Rico [Morley & Smith, 1977; White, 1976; Strassman, 1968] that firms use inappropriate production technique especially in the context of employment creation in developing countries such as Malaysia.

Several reasons have been advanced for the lack of modification of technology transfered from advanced to developing countries.

"Engineering" versus " Economic" man.

Perhaps the most convincing explanation of choice of techniques which are not compatible with cost and resource considerations is provided by Wells (1973). In a number of Indonesian manufacturing industries, Wells noted that investment decisions are strongly influenced by the objectives of engineers which are not necessarily the same as those of the "economic man". The "engineering man" prefers to solve operational problems by managing machines rather than persons; he typically aims at producing the highest quality products, and tends to believe that the most technically advanced machines in an engineering sense are also the most economically efficient ones. Firms managed by "engineering man" would then tend to use more capital-intensive, less appropriate techniques of production.

Costs and risks of technology search.

Morley & Smith (1977) suggest other factors in addition to the "engineering man" motives of private firms in their decisions on the choice of techniques. According to Morley and Smith, firms in Brazil are willing to reduce their profits by choosing an inappropriate capital-intensive technique when this choice reduces the risks of their operations. Thus, firms do not search for a lower cost technology or try to adapt to the factor proportions in the country when they perceive a higher risk of technological failure, breakdowns, delays, cost overrun, unacceptable quality, etc. More importantly, they conclude that firms in Brazil are not forced to use an appropriate technology, since the competitive environment in Brazil allowed inefficient firms to exist.

Morley and Smith's observations are of interest because the experiences in the Brazilian manufacturing sector may be similar to the conditions in Malaysia.

Position in international trade.

A particularly interesting explanation of the choice of capital-intensive technology by firms in developing countries such as Malaysia is their position in international trade. This operates through both imports and exports. If developing countries are to make real inroads into advanced country markets, they have to produce goods with acceptable quality and tastes for these markets. The exporting developing country also has to keep abreast with product and technology changes over time if it is to maintain its position in export markets. Competition in other developing countries where goods from developed countries are available make it essential to keep up with advanced country product developments. Thus even in the markets of neighboring LDCs, unless there are special trading arrangements, the exported products have to compete with developed country technology and the best way of doing so seems to be to adopt it.

For "market-oriented" firms that follow a strategy of product differentiation, branding and high advertising and selling costs, appropriate technology is relatively unimportant. Production costs for these firms are a less important component of total costs.

Role of multinational corporation.

Most empirical evidences point out that multinational firms investing in developing countries utilise advanced country techniques with little modification to the core plant or process and product, but with some variation in labor use in ancilliary activities. Branson (1975) concludes, on the basis of a survey of fifty multinational firms with automative parts manufacturing affiliates in developing countries that there is little technical adjustment in product designs or production techniques. Hughes & Poh Seng (1969) conclude that foreign firms in Singapore use capital-intensive methods. Similarly, in a study of the transfer of technology based on 338 manufacturing establishments in Malaysia (Hoffman & Tan, 1980), it is found that all MNC subsidiaries are turnkey projects which are heavily dependent on their parent companies for equipment and machine parts, professional and technical personnel, and even marketing support. Lim and Cheong (1981) also find that multinational electronic firms rely on their parent firm for capital equipment.

Although Pack (1972) and Boon (1969) provide counter evidences in Kenya and Mexico, most researchers believe that multinational firms are capital-intensive except in the electronics and textiles industries. They tend to be capital-intensive for a number of reasons. The basic reason is that the multinationals are more interested in maximizing their profits than in maximizing their output or employment [Stobaugh, 1984]. Even if investors are not fully profit-motivated, other reasons for the lack of modification of techniques by multinationals are low labor productivity, lack of local backup R&D, and the danger of having their technologies "stolen" if they are too simple. Furthermore, it is on the basis of advantages of possessing these unmodified capital-intensive technologies that they have become multinationals. The very essence of profiting from international operations lies in the ability to apply a given package to different areas with as little cost adaptation as possible. Hence there really may be little incentive for multinational firms to adopt labor-intensive techniques in developing host countries.

Differences in labor productivity.

Theoretically, when a certain technique of production is employed with labor of varying quality, it may yield a varying quantity and/or quality of output. Thus when a profit-maximizing firm chooses the optimal technique according to the relative factor prices, it must also take into account the factor productivities. Since most of the technology are transferred to developing countries from advanced countries without much modification, both advanced countries with a highly-trained labor force and developing countries with relatively low labor skills face the same set of alternative techniques. Due to their productivity differences, the firms in developing countries have the incentive to choose a more capital-intensive technique than would be expected on the basis of nominal factor prices. The incentive is especially strong if the productivity differential between the countries increases with the labor intensity of the technique chosen.

The effect of productivity differences on the choice of technique is a substitution process of different quality of labor. For the cost-minimizing firms or those concerned with quality of products would substitute capital for labor by choosing a more capital-intensive technique in order to compensate for the lower labor productivity.

The foregoing discussion of the range and density of the elasticity of substitution between capital and labor, and the choice of inappropriate capital-intensive technique of production, helps to clarify why there is low generation of employment opportunities in developing country's manufacturing sector. Choice of inappropriate technique of production by the modern manufacturing sector has often been cited as a major cause of unemployment, since excessively capital-intensive technology consumes the scarce capital resources of developing countries without generating sufficient increases in employment in the modern industrial sector. Three explanations for the choice of inappropriate technique of production have been advanced:

1. Besides the obviously labor-intensive technologies in electronic, textile and footwear industries, only a limited range of efficient technologies, which tend to be capital-intensive, are available to firms in developing countries such as Malaysia.
2. The factor prices faced by firms in developing countries may be distorted by laws concerning industrial employment, exchange rate policies, goverment tax and capital invesment incentives.
3. The decisions and the nature of the firms themselves also determine the choice of more capital-intensive techniques with minimal adaptation or modifications to the factor proportions of developing countries.

When firms in developing countries are seen to choose inappropriate techniques in their operations, and when efficient substitution possibililies are limited, the policy implications are striking both for the firms themselves and for the development policies of the developing countries. The conclusions to this study, the policy implications and recommendations for further research are discussed in the following chapter VI.

CHAPTER VI.
CONCLUSIONS AND POLICY IMPLICATION

A number of major conclusions related to the utility of the elasticity of substitution for policy making may be distilled from the foregoing research and discussion. The first concerns the usefulness of comparing estimates of the elasticity of substitution drawn from different sample bases and levels of aggregation and different equations reflecting varying assumptions about the production function. Although a generalisation of low elasticity of substitution in Malaysian manufacturing sector can be inferred from the study, strict comparability of estimates requires very strict comparability of both the treatment of data, assumptions and the estimation procedures. Nerlove (1967) and Morawetz (1973) conclude that even slight variations in the period or concepts can produce different estimates of the elasticity of substitution. It is not possible to identify industries with consistently high or low elasticities. In this study however, it is interesting to note the similarities in the range and the generally low elasticity of substitution in Malaysian manufacturing sector, either through the CES production function or the translog cost function approaches. It is however not possible to identify industries with consistently high or low elasticities and industry rankings too tend to be quite unstable. As such, it is not possible to interprete with confidence the point estimates of each industry.

Furthermore, attempts to estimate elasticities of substitution using econometric methods suffer from a number of shortcomings. Among the most important of these is the assumption of homogeneous inputs to produce a single homogeneous output. No account is taken of other factors such as the quality of management, existence of different qualities of labor and different types of capital equipment. Econometric measurements also suffer from the difficulties of incorporating technical change, working capital and varying rates of capital utilisation over time. Attempts are made in this study to overcome possible biases from aggregation problems, the effects of technical change, economies of scale and imperfections in both product and commodity markets. The results however are not satisfactory.

In light of these empirical problems, the utility of the substitution-possibility indicator could be greatly reinforced by estimates based on a better quality firm level data based on empirical surveys, field investigations and interviews with entreprenuers. Since Morawetz's criticisms of the elasticity estimated by econometric methods, there have been substantial developments and improvements in estimation procedures and econometric techniques especially concerning simultaneity problems and technical procedures for improving the quality of data for estimation purposes. Being relatively less costly, econometric measurements can still be useful.

Alternatively, detailed product-by-product or process-by-process engineering analysis studies have to be carried out to investigate the degree of factor substitutability and the extent to which the adoption of appropriate techniques can be expected to absorb employment. Such microeconomic tasks can be undertaken only for a number of products since such detailed investigations are costly and time consuming.

The second significance of this study is the role of substitution elasticities with respect to employment generation in the Malaysian manufacturing industries. In the short-term, the possibilities for substitution between capital and labor in the majority of industries appear to be rather limited. Thus, factor price policies are not likely to result in important changes in the techniques of production and employment in the Malaysian manufacturing sector. In the long-term however, the establishment of proper factor prices to reflect the true scarcities of capital and labor is very important. This has been a familiar refrain from economists over the past ten years, but it can still bear repeating. Efforts must be made to reduce the subsidies to capital use. The cumulative employment effect of setting factor prices right can be quite substantial. The new factor prices would favor the investment of new capital in more labor-intensive industries. Furthermore, where alternative technologies are available, each industry would be encouraged to use the more labor-intensive technique of production. Moreover, by increasing the price of capital relative to labor would induce an increase in the prices of capital-intensively produced goods. This will in turn result in shifting the composition of output in favor of goods with a higher employment content.

Limited factor substitution is due to the limited range of alternative technologies available. An obvious policy implication to increase employment generation is to expand the range of appropriate technologies available to the firms. This involves not only improving the information and technology networking systems, but also determining the channels and types of investments associated with the transfer of technology from industrialised countries, and improving local scientific research and technological development.

The correct choice of technique and appropriate factor proportions can positively influence employment generation in the manufacturing sector. However, changing one policy in isolation may not provide the impact on the demand for labor as envisaged. As such, other policies which affect employment need to be addressed simultaneously. These policies include the output composition or the product-mix to be manufactured, the types and scale of industries to be promoted, the increase in productivity and capacity utilisation of the manufacturing sector.

Product composition has obvious affects on the magnitude of employment opportunities which can be generated with a given level of output. The composition of output at the industry level is determined by the structure of aggregate demand consisting of demand by domestic consumers, foreign consumers in the export market, the government sector and private investors. For each of these sectors, the policy implication is to influence the demand for goods which are produced labor-intensively. Thus, efforts to increase exports of labor-intensive products such as textiles, footwear, electronics must be intensified. Furthermore, the government can influence the level of employment by directing their spending favoring labor-intensive projects rather than large-scale capital-intensive projects in public works and construction

Another area of interest which has employment implications is the intra-industry-mix. In addition to the possibility that more labor-intensive industries are encouraged, it may also be possible to encourage more labor-intensively produced goods within each industry. For example, the production of soap is more labor-intensive relative to the production of detergent, and the production of leather or canvas footwear is similarly more labor-intensive relative to the production of rubber-moulded footwear.

Another policy implication is related to the rate of capacity utilisation in the manufacturing sector. By utilising the existing capital stock more intensively, will lower the capital-labor ratio, and at the same time, increase employment in the industry. Furthermore, as capital utilisation is increased, the subsequent increased need for maintenance will add further to total employment.

Although references have been made in the literature on Malaysian industrial development, few systematic studies have been undertaken, especially in quantifying the employment effects of various policy measures.

Directions for Further Research

The issues of technological choice and employment generation are important to the Malaysian economy. This research has provided the econometric estimates of the elasticity of substitution between capital and labor and a discussion of the factors which may influence the range and density of substitution, as well as the capital-intensity bias of the manufacturing firms in Malaysia. This attempt to estimate the elasticities of substitution using econometric methods however, suffer from a number of theoretical and empirical shortcomings. These shortcomings render the estimates of the elasticity of substitution a doubtful indicator on which to base economic policy formulation. Its utility could be greatly enhanced if support for prima facie estimates could be obtained from empirical surveys, investigation of the firms and interviews with entreprenuers. Since the major shortcoming of this study is the available data, reestimation with an improved firm level data will provide more confident estimates of the elasticity of substitution. Future research to estimate the elasticity of substitution can take alternative approaches as discussed below.

1. The elasticity of substitution can be reestimated using econometric methods with improvements of the data base, as well as the model specifications. It is suggested that both the capital and labor data be reformulated as the Divisia quality indices would take account of the various qualitative aspects of capital and labor, such as capacity utilisation, the rate of depreciation of capital and educational as well as age structure of labor. Furthermore, the model can be respecified to estimate elasticity of substitution between capital and different categories of labor, i.e, skilled an unskilled labor.

2. An alternative approach in estimating the elasticity of substitution between capital and labor is to investigate technological choices at the micro level, where all of the specific determinants relevant to a given choice can be ascertained and analyzed in detail. This approach, known as engineering process analysis, permits the evaluation of alternative techniques using project appraisal methods. A principal part of the investigation is to see if there are alternative means of producing the same volume of output, that is,if more workers and fewer machines (or, usually, simpler and cheaper machines) can produce the same volume as fewer workers and more machines. This is, of course, the heart of substitutability question. Thus the appropriate technique may be identified as that which minimises production cost. Policies to increase labor use in LDC's manufacturing sector have often being discussed in the literature. Policies involving factor prices, output and industry-mix, capacity utilisation and productivity have also been discussed in the Malaysian context. However, there has been few systematic analysis to quantify the direct and indirect employment effects of the various policy instruments. Future research should concentrate on systematic analysis of policy-oriented issues. Some areas of study which merit further research are:

   i) The magnitude of factor price distortions and the problem of technical substitution and employment generation. Such studies should include the calculations of the effective rate of protection, the effective subsidy of profits, the impact of capital use subsidies and the implicit taxation of the use of labor. Such studies are

useful in determining the impact of distortionary pricing policies of the Malaysian economy, especially on employment generation.
ii) The influence of product mix and quality standards on the nature of appropriate technology and employment creation. This include studies on demand elasticities of different categories of consumers and the impact of income distribution on final demand.
iii) Finally, systematic research needs to be carried out to determine the employment impact of increasing productivity and capacity utilisation.

On the whole, therefore, our estimates do not provide a very optimistic outlook on manufacturing employment possibilities and it underscores the need for a more cogent employment-oriented industrialisation policy. Factor price policies are not likely to result in important changes in technique and employment in Malaysian manufacturing sector. The conclusion derived from the empirical estimates of the generally low capital-labor substitution possibilities is reinforced by the discussion of responsiveness of firms to factor prices in Chapter V. Nonetheless, policies to increase labor absorption through price incentives, though limited, can still be important. Getting factor, product and foreign exchange prices right is very important in an open competitive economy such as Malaysia. The practice of reducing import duties on capital goods for certain industries can result in the negative impact on labor absorption. As such more equal treatment of exports and import substitutes will ensure that countries produce according to comparative advantage.

The third and perhaps the most important policy implication of this study is that the problem of employment absorption in developing countries such as Malaysia is mainly structural in nature. The eventual resolution of the employment problem depends not only on the direct employment effects of the year-to-year choice of techniques but more importantly on the general development strategies in the country. The issues related to employment absorption are multi-faceted. Besides the choice of techniques, other aspects of importance are output and industry-mix, scale of production, the rate of population growth, the location of resources, the behavioural and cultural characteristics of households and communities, the organisation and capacity to plan and implement, international trade, capital flows and transfer of technology, the ownership and management of resources and the structure of political and economic power. Long-term labor absorption in manufacturing as well as other sectors require policies which are intended to affect these conditions. They involve technical as well as social and political conditions.

With respect to the choice of technique, the policy implication is the possibility for the government to intervene by influencing the country specific production function. In the shortrun the goverment can modify the product-mix of industries. Product composition has obvious effects on the magnitude of employment opportunities which can be generated with a given level of output. Some products require more labor per unit of output than do others and, if total costs are comparable, more labor per unit of capital employed in production. Thus, present efforts to increase exports of labor-intensive products such as textiles, footwear, electronics must be intensified. Considerations however must also be given to secondary effects of manufacturing activities. Each product requires other material and capital inputs and may itself be an input in the production of other goods, when these secondary employment effects are taken into account, product preferences may have to be reordered. Another issue is the argument that employment considerations tend to favor import-substitution over export-promotion since import-substitutes are produced to meet lower income needs. It is argued that production of such goods may on average employ more labor. This argument is less persuasive when secondary employment effects are introduced, when it is applied to intermediate goods. More careful study is necessary to determine these effects on employment in the manufacturing sector.

Domestic income distribution affects the product-mix demanded within a country. Normally, high income groups demand imported goods from building materials for their houses to their food. This aspect too requires quantitative research before quantitative effects on employment can be determined. There are however various policy tools which can be used to modify income distribution and the product-mix purchased domestically. These include progressive taxation, subsidies on essential goods and encouragement of production of non-luxury goods.

Technical strategies for improving employment in Malaysia require continuing research and empirical study. The more critical problems continue to be firstly, to modify ideal strategies to conform to the settings in which they are to be implemented. Second is to operationalise and find ways and means of transferring the knowledge and information about strategies to policy-makers so as to be implementable.

BIBLIOGRAPHY

Ady, P. D. 1971. "Private Overseas Investment and the Developing Countries." in P. Ady, ed., *Private Foreign Investment in the Developing World.* New York, Praeger.

Arrow, K., H.B. Chenery, B.S. Minhas & R. M. Solow. 1961."Capital-labor Substitution & Economic Efficiency." *Review of Economics & Statistics* 63: 225-250.

Baranson, J. 1979. *Industrial Technologies for Developing Economies.* New York, Praeger.

Barten, A. P. 1969. "Maximum Likelihood Estimation of a Complete System of Demand Equations."*European Economic Review* 1: 7-73.

Barber, C. L. 1969."The Capital-Labor Ratio in Under-Developed Areas." *Philippine Economic Journal* 8: 85-89.

Beach, C. M. & J. G. MacKinnon. 1978. "A Maximum Likelihood Procedure For Regression With Autocorrelated Errors." *Econometrica* 46: 51-58.

Behrman, J. R. 1972. "Sectoral Elasticities of Substitution between Capital & Labor in a Developing Economy; Time Series Analysis in the Case of Postwar Chile." *Econometrica* 40: 311-326.

Bhalla, A. ed., 1975. *Technology & Employment in Industry.* Geneva, ILO.

Berndt, E. R. & L. R. Christensen. 1973. "The Translog Function & the Substitution of Equipment, Structures & Labor in US Manufacturing, 1929-68." *Journal of Econometrics* 1: 81-114.

Berndt, E. R. & D. O. Wood. 1975. "Technology Prices and the Derived Demand for Energy." *Review of Economics & Statistics* 42:269- 68.

Berndt, E. R. & N. E. Savin. 1975. "Estimation and Hypothesis Testing in Singular Equation Systems with Autoregressive Disturbance." *Econometrica*, 43: 937-957.

Berndt, E. R. 1976. "Reconciling Alternative Estimates of the Elasticity of Substitution." *Review of Economics & Statistics* 43: 59-68.

Berndt E. R. & M. S. Khaled. 1979. "Parametric Productivity Measurement & Choice Among Flexible Functional Forms." *Journal of Political Economy* 87: 1220-1245.

Bigman, D. 1978. "Estimating the Rates of Factor Augmenting Technical Progress." *European Economic Review* 11: 305-317.

Binswanger, H. P. 1974. "The Measurement of Technical Change Biases with Many Factors of Production." *American Economic Review* 64: 964-976.

Boisvert, R. N. 1982. "The Translog Production Function: Its Properties and its Several Interpretations and Estimation Problems." A. E. Res. 82-83, Dept. of Agricultural Economics, Cornell University, 1982.

Boon, G. K. 1969. "Factor Intensities in Mexico with Special Reference to Manufacturing," in H. C. Bos, ed., *Towards Balanced International Growth.* Amsterdam, North Holland.

Christensen, L.R. & D.W. Jorgenson. 1969. "The Measurement of US Real Capital Input, 1929-1967." *Review of Income & Wealth* 15: 293-320.

Christensen, L. R. & D. W. Jorgenson. 1970. "US Real Product & Real Factor

Input, 1929-1967." *Review of Income & Wealth* 16: 19-50.
Christensen, L. R., D. W. Jorgensen & L. L. Lau. 1971. "Conjugate Duality & the Transcendental Logarithmic Production Function." *Econometrica* 39: 255-254.
Christensen L. R., D. W. Jorgenson & L. L. Lau. 1973. "Transcendental Logarithmic Production Frontiers." *Review of Economics & Statistics* 40: 28-45.
Claugue, C. K. 1979. "Capital Labour Substitution in Manufacturing in Underdeveloped Countries." *Econometrica* 37: 528-537.
Cochrane, D. & G. H. Orcutt. 1949. "Application of Least Squares Regression to Relationships Containing Autocorrelated Error Terms." *Journal of American Statistical Association* 44: 32-61.
Costa, E. 1973. "Maximising Employment in Labor-Intensive Development Programs." *International Labor Review* 108: 371-93.
Department of Statistics. 1964. *Census of Manufacturing Industries Peninsular Malaysia,* 1963. Kuala Lumpur, Government Printers.
Department of Statistics. 1964. *Census of Manufacturing Industries Peninsular Malaysia,* 1968. Kuala Lumpur, Government Printers.
Department of Statistics. "Survey of Manufacturing Industries Peninsular Malaysia, 1964-1976." Kuala Lumpur, Government Printers.
Department of Statistics. "Industrial Survey of Malaysia, 1978-1984." Kuala Lumpur, Government Printers.
Diewert, W. E. 1971. "An Application of the Shephard Duality Theorm: A Generalised Leontief Production Function." *Journal of Political Economy* 79: 481-507.
Diwan, R. K. 1965. "An Empirical Estimate of the Elasticity of Substitution Production Function." *The Indian Economic Journal* 12: 112-28.
Draper, N. R. and H. Smith. 1981. *Applied Regression Analysis.* New York, John Wiley & Sons.
R. S. Eckaus. 1977. *Appropriate Technologies for Developing Countries.* Washington D. C. National Academy of Sciences.
Elbadawi, I, A. R. Gallant & G. Souza, 1983. "An Elasticity can be Estimated Consistently without a Priori Knowledge of Functional Form." *Econometrica* 51: 1731-1751.
Feldstein, M. S. 1967. "Alternative Methods of Estimating a CES Production Function for Britain." *Economica* 34: 136-142.
Ferguson, C. E. 1965, "Time-Series Production Functions and Technological Progress in American Manufacturing Industry." *Journal of Political Economy:* 135-148.
Field, B. C. & C. Grebenstein. "Capital-Energy Substitution in US Manufacturing." *Review of Economics & Statistics* 62: 207-212.
Freeman, R. B. & J. L. Medoff. 1982. "Substitution between Production Labor and Other Inputs in Unionised and Nonunionised Manufacturing." *Review of Economics & Statistics* 64: 220-233.
Fong Chan Onn. 1980. "Appropriate Technology: An Empirical Study of Bicycle Manufacturing in Malaysia." *The Developing Economies* 53: 96-115.

Fuss, M.A. 1977. "The Demand for Energy in Canadian Manufacturing: An Example of the Estimation of Production Structures with Many Inputs." *Journal of Econometrics* 5: 89-116.
Gan Wee Beng. 1987. "The Ringgit Exchange Rate and the Malaysian Economy." Malaysian Economic Convention. Kuala Lumpur.
Gaude, J. 1975. "Capital-Labor Substitution Possibilities: A Review of Empirical Evidence." in A. S. Bhalla ed. *Technology and Employment in Industry.* Geneva, ILO.
Government of Malaysia. *Second Malaysia Plan, 1970-75.* Kuala Lumpur, Government printers.
Government of Malaysia. *Third Malaysia Plan, 1976-80.* Kuala Lumpur, Government Printers.
Government of Malaysia. *Midterm Review of Third Malaysia Plan, 1978.* Kuala Lumpur, Government Printers.
Government of Malaysia. *Fourth Malaysia Plan (FMP) 1980-85.* Kuala Lumpur, Government Printers.
Government of Malaysia. *Midterm Review of Fourth Malaysia Plan, 1984.* Kuala Lumpur, Government Printers.
Government of Malaysia. *Fifth Malaysia Plan, 1986-90.* Kuala Lumpur, Government Printers.
Griffin, J. M. & P. R. Gregory. 1986. "An Inter-Country Translog Model of Energy Substitution Responses." *American Economic Review* 66: 845-857.
Griliches, Z. 1967. "Production Function in Manufacturing: Some Preliminary Results." in M. Brown, ed. *The Theory and Empirical Analysis of Production.* New York, Columbia University Press.
Halvorsen, R. 1977. "Energy Substitution in US Manufacturing." *Review of Economics & Statistics* 59: 381-388.
Halvorsen R. & T.R. Smith. 1986. "Substitution Possibilities for Unpriced Natural Resources: Restricted Cost Functions for the Canadian Mining Industry." *Review of Economics & Statistics* 68: 398-405.
Hill, H. 1983. "Choice of Technique in the Indonesian Weaving Industry." *Economic Development & Cultural Change* 31: 337-354.
Hoffman, L. & S. E. Tan. 1980. *Industrial Growth, Employment and Foreign Investment in Peninsular Malaysia.* Oxford University Press, Kuala Lumpur.
Hughes, H. & P. S. You. 1969. *Foreign Investment and Industrialisation in Singapore.* Madison, Wisconsin Press.
Ioannides, Y. M. & M. Caramanis. 1979. "Capital-Labor Substitution in a Developing Country: the Case of Greece: A Note." *European Economic Review* 12: 101-110.
Jae Wan Chung. 1987. "On the Estimation of Factor Substitution in Translog Model." *Review of Economics & Statistics.* 69: 409-417.
Jae Won Kim. 1984. "CES Production Functions in Manufacturing & Problems of Industrialisation in LDCs: Evidence from Korea." *Economic Development & Cultural Change* 33: 143-165.
Jomo K. Sundram, ed., 1985. *Malaysia's New Economic Policies: Evaluations of the Mid-Term Review of The Fourth Malaysia Plan.* Malaysian Economic association, Kuala Lumpur.

Kmenta J. & R. F. Gilbert. 1968. "Small Sample Properties of Alternative Estimators of Seemingly Unrelated Regressions." *Journal of the American Statistical Association* 63: 1180-1200.

Koizumi, T. 1976. "A Further Note on Definition of Elasticity of Substitution in the Many Input Case." *Metro-Economica* 28: 152-55.

Kulatilaka, N. 1985. "The Specification of Partial Static Equilibrium Models." *The Review of Economics & Statistics* 69: 327-335.

Lancaster, K. 1966. "New Approach to Consumer Theory." *Journal of Political Economy* 74: 132-157.

Lecraw, D. J. 1979. "Choice of Technology in Low Wage Countries: A Non-neoclassical Approach." *Quanterly Journal of Economics* 93: 631-54.

Lianos, T.P. 1975. "Capital-Labor Substitution in a Developing Country: The Case of Greece." *European Economic Review* 6: 129-141.

Lianos, T. P. 1976. "Factor Augmentation in Greek Manufacturing, 1958-1969." *European Economic Review* 8: 15-31.

Lim, D. 1973. *Economic Growth and Development in West Malaysia, 1947-1970.* Kuala Lumpur, Oxford University Press.

Lim K. C. & Cheong K. C. 1981. "Transfer of Technology to Malaysia: Case Study of Electronics and Electrical Industries in Malaysia." Workshop on Negotiations for Technology Transfer Through MNC. INTAN, Kuala Lumpur.

Leontief, W. 1964. "An International Comparison of Factor Costs and Factor Use." *American Economic Review* 54: 335-45.

Lopez, R. E. 1980. "The Structure of Production and The Derived Demand For Inputs in Canadian Agriculture." *American Journal of Agricultural Economics* 62: 38-45.

Lu, L. & L. Fletcher. 1968. "A Generalisation of the CES Production Function." *Review of Economics & Statistics* 50: 449-53.

Maisom, Abdullah. 1979. "An Appraisal of The Incentive Schemes For Foreign Investment in Malaysia." *Development Forum* 10:1-20.

Malaysian Industrial Development Authority (MIDA), 1963-84. Annual Reports, 1963-1984. MIDA, Kuala Lumpur.

McFadden, D. 1963. "Further results on CES Production Functions." *Review of Economic Studies* 40: 73-83.

Morawetz, D. 1976. "Elasticities of Substitution in Industries: What do we Learn from Econometric Estimates?". *World Development* 4: 11-15.

Morley, S. A. & G.W. Smith. 1977. "The Choice of Technology: Multinalional firms in Brazil." *Economic Development & Cultural Change* 25: 239-263.

Mundlak Y. 1968. "Elasticities of Substitution and The Theory of Derived Demand." *Review of Economic Studies* 35: 225-36

Nadiri, M. I. 1970. "Some Approaches to the Theory & Measurement of Total Factor Productivity: A Survey." *Journal of Economic Literature* 8: 1137-1177.

Nerlove, M. 1967. "Recent Empirical Studies of The CES and Related Production Function." in M. Brown, ed., *The Theory and Empirical Analysis of Production.* New York, NBER.

O'Donnell, A. T. & J. K. Swales. 1979. "Factor Substitution, the CES Production Function and UK Regional Economics."*Oxford Economic Papers* 30: 460-476.

Pack, H. 1972. "The Use of Labor Intensive Techniques in Kenyan Industry" in *Technology and Economics in International Development*. Washington D. C., Agency for International Development.

Pack, H. 1984. "Productivity and Technical Choice: Applications to the Textiles Industry." *Journal of Development Economics* 16: 153-176.

Paul Chan, 1979. "MNC's and the Factor Proportions Problem: The Malaysian Experience." Seminar on Transnational Corporations, INTAN, Kuala Lumpur.

Pollack, R.A., R.C. Sickles & T.J. Wales. 1984. "The CES-Translog: Specification and Estimation of a new Cost Function." *Review of Economics & Statistics* 66 : 602-607.

Ravis, G. 1979. "Appropriate Technology in the Dual Economy: Reflections on Philippine & Taiwanese Experience." in A Robinson, ed., *Appropriate Technology for Third World Development*. New York, St. Martin's Press.

Rao, P. & Z. Griliches. "Small Sample Properties of Several Two-Stage Regression Methods in the Context of Auto-Correlated Errors." *Journal of American Statistical Association* 64: 253-272.

Ray, S. C. 1982. "A Translog Cost Function Analysis of the U.S. Agriculture, 1939-1977." *American Journal of Agricultural Economics* 64: 490-498.

Rubble, W. L. 1968. "Improving the Computation of Simultaneous Stochastic Linear Equations Estimates", Ph. D. Michigan State University.

Rushdi, A. A. 1982. "Factor Substitutability in the Manufacturing Industries of Bangladesh: An Application of the Translog Cost Model." *Bangladesh Development Studies* 10: 85-105.

Samir, A. 1969. "Levels of Renumeration, Factor Proportions and Income Differentials with Special Reference to Developing Countries." in A.D. Smith, ed., *Wage Policy Issues in Economic Development*. London, MacMillan & Co.

Schumacher, E.F. 1973. *Small is Beautiful: A Study of Economics as if People Mattered*. New York, Harper & Row.

Siggel, E. 1986. "Protection, Distortions & Investment Incentives in Zaire: A Quantitative analysis." *Journal of Development Economics* 22: 295-320.

Sicat, G. P. 1970. "Capital-Labor Substitution in Manufacturing in a Developing Economy: The Philippines." *The Developing Economies* 8: 24-38.

Stewart, F. 1972. "Choice of Technique in Developing Countries." *Journal of Development Studies* 9: 99-121.

Stewart, F. 1979. "International Technology Transfer: Issues & Policy Option" World Bank Working Paper No 344. New York.

Strassman, W.P. 1968. *"Technological Choice and Economic Development: The Manufacturing Experience of Mexico & Puerto Rico."* Ithaca, Cornell University Press.

Stobaugh, R. 1984. *Technology Crossing Borders: The Choice, Transfer and Management of International Technology Flows*. Massachusettes, HBSP.

Timmer C.P., 1984. "The Choice of Technique in Indonesia" in Stobaugh, ed., *Technology Crossing Borders*. Massachusettes, HBSP.

Toh Mun Heng. 1985. "Technical Change, Elasticity of Factor Substitution and Returns to Scale in Singapore Manufacturing Industries." *The Singapore Economic Review* 30: 36-56.

Tsao Yuan. 1985. "Growth Without Productivity: Singapore Manufacturing in the 1970's. *Journal of Development Economics* 18: 25-38.

Uzawa, H. 1962. "Production Functions with Constant Elasticities of Substitution." *Review of Economic Studies* 29 : 291-299.

Varian, H. R. 1978. *Microeconomic Analysis.* New York, W.W. Norton & Company.

Vashist D. C. 1985. "Substitution Possibilities and Price Sensitivity of Energy Demand in Indian Manufacturing." *The Indian Economic Journal* 32: 84-97.

Wells, L. T. 1984. "Economic Man & Engineering Man.: Choice of Technology in a Low Wage Country," in Stobaugh, ed., *Technology crossing Borders; International Technology Flows.* Mass. HBSP.

White, L. J. 1978. "Evidence on Appropriate Factor Proportions for Manufacturing in Less Developed Countries: A Survey." *Economic Development & Cultural Change* 27: 27-29.

Wills, J. 1979. "Technical Change in the US Primary Metals Industry." *Journal of Econometrics* 10: 85-98.

Young, K.1980. *Malaysia: Growth & Equity in a Multiracial Society.* New York, World Bank.

Zellner, A. 1962. "An Efficient Method of Estimating Seemingly Unrelated Regressions and Tests for Aggregation Bias." *Journal of the American Statistical Association* 57: 349-368.

Zellner, A. & D. S. Huang. 1962. "Further Properties of Efficient Estimators for Seemingly Unrelated Regression Equations." *International Economic Review* 3: 300-313.

APPENDIX : TIME SERIES DATA FOR ESTIMATION OF ELASTICITY OF SUBSTITUTION USING CES PRODUCTION FUNCTION AND TRANSLOG COST FUNCTION

INDUSTRY: Slaughtering, Preserving and Preparing Meat

Year	VALUE ADDED ('000)	WAGES & SALARIES ('000)	LABOR (NO OF PERSONS)	FIXED ASSETS ('000)	COST OF INPUT ('000)	DEPRE-CIATION ('000)	CPI (INDEX)
1963	–	–	–	–	–	–	–
1964	–	–	–	–	–	–	–
1965	–	–	–	–	–	–	–
1966	–	–	–	–	–	–	–
1967	–	–	–	–	–	–	–
1968	–	–	–	–	–	–	–
1969	–	–	–	–	–	–	–
1970	5216	708	546	2018	5828	171	56
1971	1326	726	503	2800	6421	238	57
1972	1623	338	277	3049	6673	271	59
1973	4139	469	344	3186	7430	270	65
1974	5661	585	489	4999	12110	424	77
1975	6788	1380	673	5162	11109	460	80
1976	7679	1893	728	8159	13138	494	82
1977	9631	1159	519	8503	13692	472	86
1978	9671	1632	309	8846	14244	446	90
1979	11801	1705	454	9500	15297	454	94
1980	13932	2460	641	11735	18896	528	100
1981	16062	3215	828	13969	20767	684	110
1982	14309	2500	502	9864	20307	641	171
1983	12555	1775	276	7988	19847	558	120
1984	19302	2391	356	7930	18852	553	126

Appendix 111

INDUSTRY: Ice Cream Manufacturing

Year	VALUE ADDED ('000)	WAGES & SALARIES ('000)	LABOR (NO OF PERSONS)	FIXED ASSETS ('000)	COST OF INPUT ('000)	DEPRE- CIATION ('000)	CPI (INDEX)
1963	2144	856	424	–	–	–	–
1964	2363	868	426	–	–	–	–
1965	2513	939	483	–	–	–	–
1966	2634	992	512	–	–	–	–
1967	2660	972	472	–	–	–	–
1968	3685	1177	490	–	–	–	–
1969	3685	1177	490	1807	2834	90	56
1970	3025	1105	423	2629	3723	131	56
1971	3299	880	375	3041	3498	152	57
1972	3883	896	358	3901	3763	351	59
1973	6176	1088	419	4190	5797	377	65
1974	6761	1250	469	4490	7994	524	77
1975	6452	1714	562	14567	10613	682	80
1976	7443	2196	524	15817	12217	1425	82
1977	11489	2486	546	15460	12403	1391	86
1978	10761	2775	568	15102	12589	1223	90
1979	11144	3201	707	19689	15059	1423	94
1980	13922	4218	845	24187	17482	1934	100
1981	16699	5234	983	28686	19905	2294	10
1982	18306	5268	676	28904	209272	812	117
1983	19912	7310	916	29462	21949	3460	120
1984	27815	6981	714	36690	29254	3502	126

INDUSTRY: Manufacture of Other Dairy Products

Year	VALUE ADDED ('000)	WAGES & SALARIES ('000)	LABOR (NO OF PERSONS)	FIXED ASSETS ('000)	COST OF INPUT ('000)	DEPRE- CIATION ('000)	CPI (INDEX)
1963	4738	941	241	–	–	–	–
1964	6171	987	365	–	–	–	–
1965	10444	1180	456	–	–	–	–
1966	19175	1638	544	–	–	–	–
1967	17589	1991	618	–	–	–	–
1968	21856	2736	618	–	–	–	–
1969	21856	2736	622	15907	66527	2863	56
1970	20912	2836	633	19063	72298	3431	56
1971	12026	3291	718	24068	84306	4332	57
1972	15106	3859	826	23177	124255	4171	59
1973	22502	4351	824	22936	149995	4128	65
1974	27836	5694	986	26048	186579	4688	77
1975	33910	6585	966	26687	233967	5819	80
1976	69330	8696	1171	30549	246264	5732	82
1977	70835	9991	1316	40125	265718	6222	86
1978	72340	11285	1461	49700	285170	7033	90
1979	71968	12488	1534	60817	325001	8982	94
1980	100138	16330	1762	91407	395577	11711	100
1981	128308	20171	1990	121996	466153	13419	110
1982	135049	20960	1977	126441	465450	13908	117
1983	141789	22356	1767	152082	464746	15848	120
1984	153178	25243	1912	173597	450759	18090	26

INDUSTRY: Pineapple Canning

Year	VALUE ADDED ('000)	WAGES & SALARIES ('000)	LABOR (NO OF PERSONS)	FIXED ASSETS ('000)	COST OF INPUT ('000)	DEPRE-CIATION ('000)	CPI (INDEX)
1963	7995	4077	1548	–	–	–	–
1964	9461	4621	1922	–	–	–	–
1965	15854	5587	2210	–	–	–	–
1966	13769	5804	2391	–	–	–	–
1967	13952	6377	2662	–	–	–	–
1968	13200	6445	2760	–	–	–	–
1969	13200	6445	2760	7704	36096	385	56
1970	12740	6253	2758	7397	38125	369	56
1971	10163	5975	2500	12669	37010	633	57
1972	10243	5535	3028	14119	36001	705	59
1973	10324	5367	2995	13538	39209	676	65
1974	12076	5629	2765	21252	50592	862	77
1975	15339	5891	2615	17509	39098	927	80
1976	18666	7180	2508	17813	42735	751	82
1977	18636	7080	2415	17886	46369	894	86
1978	18605	7234	2321	18758	50003	1127	90
1979	18878	7387	2173	17484	48581	855	94
1980	14272	7167	2028	15132	45451	756	100
1981	9666	7321	1883	12780	42520	639	110
1982	11090	7474	1809	17634	43967	881	117
1983	12513	7414	1609	16047	45413	943	120
1984	17666	6989	1518	16026	44542	941	126

INDUSTRY: Canning of Vegetables, Pickles, etc (others)

Year	VALUE ADDED ('000)	WAGES & SALARIES ('000)	LABOR (NO OF PERSONS)	FIXED ASSETS ('000)	COST OF INPUT ('000)	DEPRE- CIATION ('000)	CPI (INDEX)
1963	–	–	–	–	–	–	–
1964	–	–	–	–	–	–	–
1965	–	–	–	–	–	–	–
1966	–	–	–	–	–	–	–
1967	–	–	–	–	–	–	–
1968	–	–	–	–	–	–	–
1969	–	–	–	–	–	–	–
1970	6234	2443	1602	6947	26276	1389	56
1971	6462	3062	1998	9132	34540	1826	57
1972	15331	3692	2503	11317	42804	2263	59
1973	25853	7456	5917	20198	79922	4039	65
1974	21817	7972	4392	23467	68401	4693	77
1975	31797	4879	3086	23027	64672	5819	80
1976	42669	12344	5921	28985	98694	5732	82
1977	69816	10339	4916	27297	81891	5459	86
1978	96962	18333	3910	25609	89631	7033	90
1979	79250	19365	8761	24939	164817	8982	94
1980	61537	23811	8760	72221	216663	10110	100
1981	43825	28256	8758	89502	268506	12530	110
1982	57324	22035	5932	99461	248652	13924	117
1983	70822	21751	5611	726325	304896	15849	120
1984	67395	22546	5246	75707	227121	11356	126

INDUSTRY: Coconut Oil Manufacturing

Year	VALUE ADDED ('000)	WAGES & SALARIES ('000)	LABOR (NO OF PERSONS)	FIXED ASSETS ('000)	COST OF INPUT ('000)	DEPRE-CIATION ('000)	CPI (INDEX)
1963	4750	1369	824	–	–	–	–
1964	4379	1472	719	–	–	–	–
1965	7122	1574	912	–	–	–	–
1966	8516	1754	935	–	–	–	–
1967	7262	1774	979	–	–	–	–
1968	8629	1793	876	–	–	–	–
1969	12553	2341	1170	43280	114728	3246	56
1970	10254	2854	1321	86840	141358	6513	56
1971	10483	2886	1410	40071	133195	3005	57
1972	12399	3291	1568	32740	109230	2455	59
1973	22144	3695	1534	43509	166204	3263	65
1974	27372	3478	1499	47560	280382	3567	77
1975	8622	3261	1378	13839	132122	1058	80
1976	5615	3268	1328	13332	138151	1196	82
1977	9763	2969	1155	13161	147841	1107	86
1978	13911	2670	1182	12994	124231	1956	90
1979	14729	2544	1146	14723	154878	1026	94
1980	12696	2991	1178	23191	132674	1623	100
1981	10664	3692	1409	31662	140470	2375	110
1982	7912	3991	1266	15715	112010	1100	117
1983	5160	2505	1200	15789	107391	1086	120
1984	7667	2334	1155	15864	119403	1110	126

INDUSTRY: Palm Oil Manufacturing

Year	VALUE ADDED ('000)	WAGES & SALARIES ('000)	LABOR (NO OF PERSONS)	FIXED ASSETS ('000)	COST OF INPUT ('000)	DEPRE-CIATION ('000)	CPI (INDEX)
1963	–	–	–	–	–	–	–
1964	–	–	–	–	–	–	–
1965	–	–	–	–	–	–	–
1966	–	–	–	–	–	–	–
1967	–	–	–	–	–	–	–
1968	–	–	–	–	–	–	–
1969	–	–	–	–	–	–	–
1970	–	–	–	–	–	–	–
1971	–	–	–	–	–	–	–
1972	–	–	–	–	–	–	–
1973	81013	718	3298	113600	180180	10224	65
1974	198423	11532	4418	147522	427833	13276	77
1975	301578	17009	5206	155933	543956	14093	80
1976	303920	23930	7006	250381	822904	21518	82
1977	412417	29389	7970	324764	1110920	29228	86
1978	520913	34848	8933	399146	1365320	32029	90
1979	560720	51749	11260	579894	1979714	51259	94
1980	600528	72370	13517	753698	2855165	67832	100
1981	640335	92991	15774	927501	3511888	83475	110
1982	566626	107288	16473	1152785	4059893	103750	117
1983	492916	102201	15013	1190494	4607898	127151	120
1984	927048	113896	14472	1240060	6308760	129400	126

INDUSTRY: Palm Kernel Oil Manufacturing

Year	VALUE ADDED ('000)	WAGES & SALARIES ('000)	LABOR (NO OF PERSONS)	FIXED ASSETS ('000)	COST OF INPUT ('000)	DEPRE-CIATION ('000)	CPI (INDEX)
1963	–	–	–	–	–	–	–
1964	–	–	–	–	–	–	–
1965	–	–	–	–	–	–	–
1966	–	–	–	–	–	–	–
1967	–	–	–	–	–	–	–
1968	–	–	–	–	–	–	–
1969	–	–	–	–	–	–	–
1970	–	–	–	–	–	–	–
1971	–	–	–	–	–	–	–
1972	–	–	–	–	–	–	–
1973	6641	600	319	5423	21550	352	65
1974	9473	1375	609	13333	103468	866	77
1975	6107	2187	743	18827	83669	1230	80
1976	11920	2522	729	16898	95316	1146	82
1977	16546	3707	963	21915	123614	1424	86
1978	21171	4891	1196	26932	151911	1826	90
1979	34149	6489	1296	25347	146271	2900	94
1980	47129	8342	1550	46021	195012	2991	100
1981	60105	10194	1803	66694	354051	3335	110
1982	56665	13553	1759	74309	465945	3115	117
1983	53227	11353	1772	75993	577838	3917	120
1984	80582	14185	1628	88121	677102	4041	126

INDUSTRY: Manufacture of Vegetable Oils and Animal Fats

Year	VALUE ADDED ('000)	WAGES & SALARIES ('000)	LABOR (NO OF PERSONS)	FIXED ASSETS ('000)	COST OF INPUT ('000)	DEPRE-CIATION ('000)	CPI (INDEX)
1963	634	150	126	–	–	–	–
1964	659	172	120	–	–	–	–
1965	687	171	123	–	–	–	–
1966	602	186	167	–	–	–	–
1967	623	190	148	–	–	–	–
1968	647	209	140	–	–	–	–
1969	731	207	161	544	2414	48	56
1970	626	219	178	579	2494	52	56
1971	580	256	183	573	2611	52	57
1972	913	355	233	959	3853	86	59
1973	9726	1759	436	6006	20542	540	65
1974	11008	2044	438	5961	30827	536	77
1975	15975	1313	313	6528	30946	619	80
1976	20092	4461	927	24448	107727	1509	82
1977	27854	5572	918	25910	128809	1813	86
1978	35615	6683	908	27371	149891	2134	90
1979	41894	10185	1137	38379	230226	2900	94
1980	38085	10767	1202	52466	258068	3934	100
1981	34276	11348	1267	66553	285909	4591	110
1982	33766	14437	1294	59596	278576	4759	117
1983	33255	7479	733	49313	271243	3917	120
1984	54866	8730	677	39682	249837	3551	126

INDUSTRY: Rice Milling

Year	VALUE ADDED ('000)	WAGES & SALARIES ('000)	LABOR (NO OF PERSONS)	FIXED ASSETS ('000)	COST OF INPUT ('000)	DEPRE-CIATION ('000)	CPI (INDEX)
1963	8651	3400	1823	–	–	–	–
1964	8650	3428	741	–	–	–	–
1965	11880	4041	2101	–	–	–	–
1966	12594	4167	2202	–	–	–	–
1967	11884	3853	1983	–	–	–	–
1968	5419	3893	2113	–	–	–	–
1969	7733	3936	2134	15212	123498	912	56
1970	18019	4445	2330	15652	146742	939	56
1971	22171	4953	2692	17391	164644	1043	57
1972	18362	4852	2661	15413	160690	924	59
1973	31471	5841	2655	20398	214495	1223	65
1974	26422	6636	2925	33250	269299	1995	77
1975	37099	7292	3095	31786	292184	1978	80
1976	39100	7142	2808	31740	283448	1884	82
1977	34231	7351	2664	31498	267194	1889	86
1978	29361	7560	2520	37256	250940	2074	90
1979	37721	11140	3266	70290	323749	4477	94
1980	53450	18107	3994	130509	397439	7830	100
1981	69178	25074	4722	190727	471129	7940	110
1982	51915	21562	5103	124693	120002	7481	117
1983	34652	17737	3679	119396	288874	6840	120
1984	30075	17619	3638	131999	259702	7562	126

INDUSTRY: Biscuit Factories

Year	VALUE ADDED ('000)	WAGES & SALARIES ('000)	LABOR (NO OF PERSONS)	FIXED ASSETS ('000)	COST OF INPUT ('000)	DEPRE-CIATION ('000)	CPI (INDEX)
1963	6779	2924	2132	–	–	–	–
1964	6881	2963	2057	–	–	–	–
1965	6921	2831	1992	–	–	–	–
1966	7594	3152	2207	–	–	–	–
1967	8453	3291	2199	–	–	–	–
1968	8276	3449	2262	–	–	–	–
1969	8590	3690	2598	4436	27679	399	56
1970	7539	3834	2516	6036	27452	543	56
1971	7805	3905	2771	6273	28467	564	57
1972	8854	4093	2712	10156	30790	914	59
1973	12297	5011	3261	11344	45176	1020	65
1974	13992	5762	2941	11440	55467	1029	77
1975	13773	5860	3012	10866	48825	1089	80
1976	15103	7749	3574	19522	57046	1673	82
1977	22639	8369	3577	20077	60753	1805	86
1978	20127	8988	3580	20631	64460	1993	90
1979	23069	10490	3670	24398	68714	2317	94
1980	31538	14242	4809	41057	85426	2668	100
1981	40007	17993	5948	57715	102138	3751	110
1982	48914	17136	4741	64034	159349	4162	117
1983	45945	18452	4726	69374	96558	4534	120
1984	37673	18463	4033	67585	95679	4417	126

INDUSTRY: Sugar Refineries and Factories

Year	VALUE ADDED ('000)	WAGES & SALARIES ('000)	LABOR (NO OF PERSONS)	FIXED ASSETS ('000)	COST OF INPUT ('000)	DEPRE-CIATION ('000)	CPI (INDEX)
1963	–	–	–	–	–	–	–
1964	–	–	–	–	–	–	
1965	–	–	–	–	–	–	–
1966	–	–	–	–	–	–	–
1967	–	–	–	–	–	–	–
1968	21866	3513	698	–	–	–	–
1969	31118	4346	964	–	–	–	–
1970	66845	7475	1790	–	–	–	–
1971	68024	7323	1856	72230	131697	7940	57
1972	34378	10372	2695	55000	139578	6050	59
1973	28767	5600	1965	81602	207085	8976	65
1974	25757	4521	1619	85844	251312	9442	77
1975	45072	11858	3414	127605	331790	13822	80
1976	58807	11563	2658	147004	376900	11851	82
1977	72039	11858	2484	147748	378807	11852	86
1978	85271	12153	2308	148491	380709	11891	90
1979	81669	12951	1801	148465	380645	10459	94
1980	78066	15285	2199	146575	385491	11726	100
1981	74464	17619	2597	132304	334133	14553	110
1982	102612	20689	2788	144185	570089	16534	117
1983	130774	24398	2892	147608	506045	17481	120
1984	123433	25796	3256	140492	501264	17454	126

INDUSTRY: Manufacture of Cocoa, Chocolate and Sugar Confectionery

Year	VALUE ADDED ('000)	WAGES & SALARIES ('000)	LABOR (NO OF PERSONS)	FIXED ASSETS ('000)	COST OF INPUT ('000)	DEPRE- CIATION ('000)	CPI (INDEX)
1963	–	–	–	–	–	–	–
1964	–	–	–	–	–	–	–
1965	–	–	–	–	–	–	–
1966	–	–	–	–	–	–	–
1967	–	–	–	–	–	–	–
1968	–	–	–	–	–	–	–
1969	–	–	–	–	–	–	–
1970	3343	1069	712	3144	9533	230	56
1971	7064	1496	916	4463	13662	330	57
1972	6861	1832	1170	9087	22625	672	59
1973	10669	2519	1518	12890	30597	953	65
1974	13429	3006	1633	16972	37309	1255	77
1975	7235	3382	1744	20202	43377	1418	80
1976	19497	5933	2341	32734	59665	2407	82
1977	26793	7001	2488	38349	69899	2837	86
1978	34088	8069	2634	43963	80132	3667	90
1979	34139	9009	2897	46658	85044	3960	94
1980	34191	11453	3178	54450	99246	4029	100
1981	34242	13896	3459	62242	127705	4605	110
1982	46392	14230	3455	87780	142846	6495	117
1983	58541	13991	3192	90183	155989	6658	120
1984	66847	19889	3522	119194	205442	8343	126

INDUSTRY: Ice Factories

Year	VALUE ADDED ('000)	WAGES & SALARIES ('000)	LABOR (NO OF PERSONS)	FIXED ASSETS ('000)	COST OF INPUT ('000)	DEPRE-CIATION ('000)	CPI (INDEX)
1963	4433	1357	563	–	–	–	–
1964	4571	1407	596	–	–	–	–
1965	4485	1456	547	–	–	–	–
1966	4569	1323	533	–	–	–	–
1967	4493	1417	583	–	–	–	–
1968	5418	1370	557	–	–	–	–
1969	5799	1454	609	4436	3104	266	56
1970	4720	1386	589	10594	7767	355	56
1971	4807	1302	546	9731	3124	562	57
1972	5419	1333	540	9536	3324	572	59
1973	6016	1558	636	14162	4194	649	65
1974	6276	1823	636	18162	4819	697	77
1975	4520	1232	513	9917	4372	684	80
1976	3866	1400	495	10246	5622	725	82
1977	4966	1570	504	13119	5803	755	86
1978	6065	1740	512	15991	5983	823	90
1979	8170	2253	644	23581	8204	1152	94
1980	8447	3244	818	28051	10534	1402	100
1981	8724	4235	991	32520	12863	1626	110
1982	9584	3723	751	27400	9584	1830	117
1983	10440	4342	806	29895	12504	2556	120
1984	11303	4961	849	32390	15424	2591	126

124 Substitutability in Malaysian Manufacturing

INDUSTRY: Coffee Factories

Year	VALUE ADDED ('000)	WAGES & SALARIES ('000)	LABOR (NO OF PERSONS)	FIXED ASSETS ('000)	COST OF INPUT ('000)	DEPRE- CIATION ('000)	CPI (INDEX)
1963	–	–	–	–	–	–	–
1964	–	–	–	–	–	–	–
1965	–	–	–	–	–	–	–
1966	–	–	–	–	–	–	–
1967	–	–	–	–	–	–	–
1968	–	–	–	–	–	–	–
1969	–	–	–	–	–	–	–
1970	2397	781	564	1492	12092	134	56
1971	2561	796	592	1497	12151	135	57
1972	2708	802	580	1636	14701	147	59
1973	4243	1214	843	1724	16427	155	65
1974	3070	951	620	1812	18153	163	77
1975	3920	1219	769	1907	19079	166	80
1976	3881	1376	827	1901	23370	209	82
1977	4479	1506	780	2332	28692	210	86
1978	5078	1635	733	2767	34015	328	90
1979	7109	1959	863	3517	37234	292	94
1980	9139	2646	1114	6189	39284	395	100
1981	11170	3333	1364	8860	41334	408	110
1982	11145	3140	1065	7888	36768	473	117
1983	11120	3975	1081	7967	35816	596	120
1984	11787	3845	996	8574	43764	685	126

INDUSTRY: Meehoon, Noodles and Related Industries

Year	VALUE ADDED ('000)	WAGES & SALARIES ('000)	LABOR (NO OF PERSONS)	FIXED ASSETS ('000)	COST OF INPUT ('000)	DEPRE- CIATION ('000)	CPI (INDEX)
1963	–	–	–	–	–	–	–
1964	–	–	–	–	–	–	–
1965	–	–	–	–	–	–	–
1966	–	–	–	–	–	–	–
1967	–	–	–	–	–	–	–
1968	–	–	–	–	–	–	–
1969	–	–	–	–	–	–	–
1970	1644	795	703	1124	4623	89	56
1971	2072	1795	754	1859	7070	148	57
1972	3558	1175	905	3350	11836	268	59
1973	6134	2042	1538	7563	24664	305	65
1974	5060	2162	1395	8046	25465	443	77
1975	8552	2469	1523	7121	23805	548	80
1976	7229	3375	2011	8913	30749	750	82
1977	10615	3738	1884	10267	35420	821	86
1978	14000	4101	1757	11620	40088	994	90
1979	18985	5545	2371	17856	61602	1490	94
1980	23969	7686	3164	29924	74326	1795	100
1981	28954	9827	3957	41992	87049	2519	110
1982	29455	8183	2243	23260	68185	2760	117
1983	29956	10605	2815	34264	77176	3331	120
1984	39933	13986	3229	44755	99734	3580	126

INDUSTRY: Animal Feeds Manufacturing

Year	VALUE ADDED ('000)	WAGES & SALARIES ('000)	LABOR (NO OF PERSONS)	FIXED ASSETS ('000)	COST OF INPUT ('000)	DEPRE-CIATION ('000)	CPI (INDEX)
1963	2876	720	314	–	–	–	–
1964	4158	848	361	–	–	–	–
1965	5439	976	408	–	–	–	–
1966	4295	1114	512	–	–	–	–
1967	5890	1313	572	–	–	–	–
1968	8178	1966	771	–	–	–	–
1969	9982	2286	924	5350	72140	482	56
1970	10551	2420	1071	6864	85487	618	56
1971	9807	2450	1007	6016	83536	541	57
1972	11621	2479	1274	8246	90446	742	59
1973	16604	3588	1614	9856	126409	887	65
1974	19927	4003	1296	14715	156171	1324	77
1975	20301	4930	1592	21347	174964	1862	80
1976	26058	6241	1815	24757	216393	2306	82
1977	32682	7575	1918	32855	232371	2468	86
1978	39306	8909	2020	40947	248349	2503	90
1979	38934	11045	2372	48198	317877	4012	94
1980	58850	14807	2885	63511	401058	4475	100
1981	78766	18568	2671	68824	484239	4774	110
1982	76350	15595	2457	68206	451229	4774	117
1983	73934	18816	2503	105065	540295	7325	120
1984	83221	9182	2448	108420	548255	7559	126

INDUSTRY: Soft Drinks and Carbonated Beverages

Year	VALUE ADDED ('000)	WAGES & SALARIES ('000)	LABOR (NO OF PERSONS)	FIXED ASSETS ('000)	COST OF INPUT ('000)	DEPRE-CIATION ('000)	CPI (INDEX)
1963	12557	4210	2100	–	–	–	–
1964	12029	4382	2026	–	–	–	–
1965	9644	4239	1839	–	–	–	–
1966	12075	4587	1909	–	–	–	–
1967	14758	4902	1797	–	–	–	–
1968	16237	4411	1797	–	–	–	–
1969	16559	4066	1982	11149	17809	557	56
1970	17278	5062	2107	11872	19049	593	56
1971	19435	5665	1957	13747	19613	687	57
1972	22151	5150	2051	13743	28887	687	59
1973	24429	5727	2054	12783	33383	639	65
1974	26390	5760	1962	20240	36199	1012	77
1975	24012	7152	2270	20019	43916	1254	80
1976	26723	7259	2100	26038	44139	1514	82
1977	38120	9278	2464	34175	58230	1708	86
1978	49516	11296	2828	42312	72321	2187	90
1979	61855	13450	3202	44045	84535	2560	94
1980	77877	18306	3804	87442	103135	4372	100
1981	93899	23161	4405	83083	121735	4404	110
1982	98357	21175	3190	89283	126502	4464	117
1983	102815	23257	2724	91806	131269	7301	120
1984	95519	25062	2871	88039	120841	6162	126

INDUSTRY: Tobaco Manufacturing

Year	VALUE ADDED ('000)	WAGES & SALARIES ('000)	LABOR (NO OF PERSONS)	FIXED ASSETS ('000)	COST OF INPUT ('000)	DEPRE-CIATION ('000)	CPI (INDEX)
1963	28146	5739	3925	–	–	–	–
1964	37144	6034	3845	–	–	–	–
1965	40057	6322	3755	–	–	–	–
1966	40057	6322	3755	–	–	–	–
1967	62460	6215	3861	–	–	–	–
1968	57106	10034	4054	–	–	–	–
1969	79193	11353	3982	30036	188183	2553	56
1970	84566	11647	4056	35405	190138	3009	56
1971	90004	12240	4488	36686	199146	3118	57
1972	90385	12922	4249	37065	204530	3151	59
1973	130418	15965	7046	43340	213184	3684	65
1974	113878	19637	7028	56352	262313	4789	77
1975	98698	23125	5931	62228	313204	5486	80
1976	129619	23125	5931	73474	320100	6945	82
1977	138564	25648	5731	82894	359095	7046	86
1978	147509	27146	5802	92315	398089	8224	90
1979	174563	27249	6717	125043	401615	9679	94
1980	222980	35369	9190	149744	430970	12728	100
1981	271397	43488	11663	174444	460324	14827	110
1982	327425	48196	11462	205844	534938	17497	117
1983	383453	43230	5457	219988	609552	18255	120
1984	351503	45236	3840	302301	627135	25695	126

INDUSTRY: Manufacture of Leather, Leather Products and Substitutes

Year	VALUE ADDED ('000)	WAGES & SALARIES ('000)	LABOR (NO OF PERSONS)	FIXED ASSETS ('000)	COST OF INPUT ('000)	DEPRE- CIATION ('000)	CPI (INDEX)
1963	–	–	–	–	–	–	–
1964	–	–	–	–	–	–	–
1965	–	–	–	–	–	–	–
1966	–	–	–	–	–	–	–
1967	–	–	–	–	–	–	–
1968	–	–	–	–	–	–	–
1969	–	–	–	–	–	–	–
1970	1607	448	113	2296	3606	149	56
1971	904	439	106	3025	5283	196	57
1972	1040	1057	122	3304	6683	214	59
1973	1819	595	137	1193	5335	177	65
1974	1833	622	283	1136	5093	173	77
1975	943	623	359	799	3166	153	80
1976	1869	637	485	1480	4777	132	82
1977	2100	857	542	1317	4251	187	86
1978	2331	1146	598	1266	4087	190	90
1979	3045	1281	561	1463	4723	193	94
1980	3760	1536	633	1935	6251	212	100
1981	4474	1853	705	2406	7677	187	110
1982	3996	1597	490	2663	9149	207	117
1983	3517	1361	380	2587	8621	201	120
1984	3205	1566	417	4201	9870	326	126

INDUSTRY: Sawmilling

Year	VALUE ADDED ('000)	WAGES & SALARIES ('000)	LABOR (NO OF PERSONS)	FIXED ASSETS ('000)	COST OF INPUT ('000)	DEPRE- CIATION ('000)	CPI (INDEX)
1963	40202	20369	9717	–	–	–	–
1964	48402	23147	10349	–	–	–	–
1965	49602	23835	10448	–	–	–	–
1966	48983	24073	10681	–	–	–	–
1967	51523	25876	11459	–	–	–	–
1968	68757	32379	13627	–	–	–	–
1969	87263	34514	15197	50947	139991	6623	56
1970	88944	39377	16848	61810	164836	8035	56
1971	81472	39639	17357	65321	164905	8492	57
1972	120079	47528	17680	79513	238246	10336	59
1973	192124	61702	22257	113989	369461	14818	65
1974	174237	67783	22673	149108	375028	19384	77
1975	173421	73515	23581	186128	391998	25220	80
1976	232826	84389	24104	195986	570965	24881	92
1977	282594	97371	24961	206972	628108	26906	86
1978	332362	110352	25818	237957	685251	31064	90
1979	427282	145017	29692	285154	953008	36771	94
1980	423946	198880	37009	430685	916952	55989	100
1981	420610	252743	44326	380896	976215	49516	110
1982	361374	195539	31044	385991	813655	50178	117
1983	302137	143751	21807	310900	746413	35619	120
1984	290038	143772	20671	277283	594887	36046	126

INDUSTRY: Planing Mills and Joinery Works

Year	VALUE ADDED ('000)	WAGES & SALARIES ('000)	LABOR (NO OF PERSONS)	FIXED ASSETS ('000)	COST OF INPUT ('000)	DEPRE-CIATION ('000)	CPI (INDEX)
1963	2804	1793	799	–	–	–	–
1964	3493	21590	1018	–	–	–	–
1965	3536	2092	1027	–	–	–	–
1966	3257	1936	962	–	–	–	–
1967	2264	1988	987	–	–	–	–
1968	3249	1902	937	–	–	–	–
1969	3240	1905	997	2034	5222	172	56
1970	2968	2159	1179	3662	6283	311	56
1971	4794	2503	1359	3826	8640	325	57
1972	9229	3523	1811	11175	13816	949	59
1973	15106	5285	2694	21810	24466	1853	65
1974	18394	6878	2876	25212	33362	2143	77
1975	19050	7166	3071	22991	33543	2468	80
1976	20762	8379	3149	23401	40517	1788	82
1977	31002	11877	3806	32422	57176	2755	86
1978	41242	15372	4663	41442	73834	3924	90
1979	51460	17544	4891	54697	104086	4872	94
1980	54226	23455	4878	76913	117968	6537	100
1981	56991	29365	6865	99128	131849	8425	110
1982	54621	29254	5392	81372	121169	6916	117
1983	52251	23460	4402	76487	110489	5115	120
1984	61485	26687	4504	86279	123840	7333	126

132 Substitutability in Malaysian Manufacturing

INDUSTRY: Manufacture of Furniture and Fixtures

Year	VALUE ADDED ('000)	WAGES & SALARIES ('000)	LABOR (NO OF PERSONS)	FIXED ASSETS ('000)	COST OF INPUT ('000)	DEPRE-CIATION ('000)	CPI (INDEX)
1963	–	–	–	–	–	–	–
1964	–	–	–	–	–	–	–
1965	–	–	–	–	–	–	–
1966	–	–	–	–	–	–	–
1967	–	–	–	–	–	–	–
1968	–	–	–	–	–	–	–
1969	–	–	–	–	–	–	–
1970	9090	4396	2554	5141	14996	514	56
1971	10278	4246	2685	6921	20188	692	57
1972	10364	5237	3050	7056	24771	705	59
1973	19282	9556	4874	14064	48651	1406	65
1974	19849	8926	3784	13126	37694	1312	77
1975	24330	10983	4680	13827	38588	2113	80
1976	23126	11339	4831	25796	36735	2003	82
1977	29909	14434	6166	33651	47920	2692	86
1978	36692	17528	7501	41506	59105	3011	90
1979	53121	22679	7199	43976	65622	3512	94
1980	69549	35579	10577	58367	120272	4669	100
1981	85978	48478	13954	72757	150162	5820	110
1982	72753	40447	13928	67076	120396	5366	117
1983	59527	29658	16290	57182	90629	5283	120
1984	66165	36605	16811	71798	99034	5743	126

INDUSTRY: Clothing Factories

Year	VALUE ADDED ('000)	WAGES & SALARIES ('000)	LABOR (NO OF PERSONS)	FIXED ASSETS ('000)	COST OF INPUT ('000)	DEPRE-CIATION ('000)	CPI (INDEX)
1963	–	–	–	–	–	–	–
1964	–	–	–	–	–	–	–
1965	–	–	–	–	–	–	–
1966	–	–	–	–	–	–	–
1967	–	–	–	–	–	–	–
1968	–	–	–	–	–	–	–
1969	–	–	–	–	–	–	–
1970	8558	3828	3652	3547	12464	283	56
1971	12492	5041	4898	5497	27682	439	57
1972	16063	7741	7719	8164	42581	653	59
1973	28767	11003	9941	18118	74566	1449	65
1974	31547	15447	10617	21774	89427	1741	77
1975	167687	17054	11070	29584	93578	2367	80
1976	44551	21194	12039	35126	112236	2880	82
1977	57134	26531	13433	41028	131094	3282	86
1978	69716	31868	14826	46929	149948	4507	90
1979	96528	40151	16399	55450	177174	5291	94
1980	123340	59293	20919	77835	248697	6226	100
1981	150152	78434	25438	100220	318174	8017	110
1982	167807	81238	24308	106757	336054	8040	117
1983	185461	105219	26853	139139	353934	8183	120
1984	230539	129342	30200	174026	422018	8701	126

134 Substitutability in Malaysian Manufacturing

INDUSTRY: Manufacture of Paper and Paper Products

Year	VALUE ADDED ('000)	WAGES & SALARIES ('000)	LABOR (NO OF PERSONS)	FIXED ASSETS ('000)	COST OF INPUT ('000)	DEPRE- CIATION ('000)	CPI (INDEX)
1963	–	–	–	–	–	–	–
1964	–	–	–	–	–	–	–
1965	–	–	–	–	–	–	–
1966	–	–	–	–	–	–	–
1967	–	–	–	–	–	–	–
1968	–	–	–	–	–	–	–
1969	–	–	–	–	–	–	–
1970	9495	3244	1741	14818	13640	148	56
1971	11493	3798	1545	29516	27169	295	57
1972	12806	4242	2174	22378	35415	223	59
1973	17971	5645	3542	27281	44824	272	65
1974	27001	6750	3003	32693	62030	326	77
1975	32186	8598	8260	32940	62463	355	80
1976	110733	10445	4014	47334	74820	493	82
1977	78694	13165	5318	57077	90220	473	86
1978	46655	15885	4777	66818	105628	652	90
1979	62111	20095	5859	77032	121772	1603	94
1980	77568	24437	6291	96745	152934	967	100
1981	93024	28779	6722	116458	218058	1164	110
1982	61041	32623	6451	147852	215671	1478	117
1983	110314	36210	6496	206515	213283	2119	120
1984	123273	43803	6784	241532	235538	2415	126

INDUSTRY: Printing, Publishing and Allied Industries

Year	VALUE ADDED ('000)	WAGES & SALARIES ('000)	LABOR (NO OF PERSONS)	FIXED ASSETS ('000)	COST OF INPUT ('000)	DEPRE-CIATION ('000)	CPI (INDEX)
1963	15194	4381	1107	–	–	–	–
1964	3157	6001	1176	–	–	–	–
1965	574	8620	1209	–	–	–	–
1966	15023	11445	3180	–	–	–	–
1967	23111	13125	3457	–	–	–	–
1968	52558	23140	9510	–	–	–	–
1969	59255	25179	10237	48368	62422	5320	56
1970	73140	27017	11211	53868	71443	5925	56
1971	78409	30020	11440	55440	74481	6098	57
1972	96136	32576	12221	66364	94572	7300	59
1973	114551	38930	13886	88113	123413	9692	65
1974	142242	44293	14181	106475	158524	10112	77
1975	125832	41375	13674	103838	145066	10718	80
1976	133613	54366	14397	125072	176871	12242	82
1977	170201	66264	15124	141438	199169	15558	86
1978	206789	78161	15851	157803	221467	17866	90
1979	240976	94664	17730	197708	285406	18442	94
1980	314265	121983	21803	272956	358107	19566	100
1981	387553	149301	25876	348203	430808	20892	110
1982	394713	158954	20616	343318	423365	22599	117
1983	401873	162674	16752	301220	415921	25255	120
1984	495053	180092	18519	342184	420380	28688	126

INDUSTRY: Industrial Chemicals

Year	VALUE ADDED ('000)	WAGES & SALARIES ('000)	LABOR (NO OF PERSONS)	FIXED ASSETS ('000)	COST OF INPUT ('000)	DEPRE-CIATION ('000)	CPI (INDEX)
1963	1491	394	415	–	–	–	–
1964	1640	413	380	–	–	–	–
1965	1641	508	474	–	–	–	–
1966	2218	644	530	–	–	–	–
1967	2889	770	538	–	–	–	–
1968	3076	876	538	–	–	–	–
1969	3833	1140	752	–	–	–	–
1970	7572	1380	980	–	–	–	–
1971	6633	1638	1839	15126	10672	1210	57
1972	12725	3414	2025	35383	18777	2830	59
1973	19549	5750	2158	69608	32023	5568	65
1974	34844	5375	2362	95344	48355	7627	77
1975	50314	9370	2717	111637	65636	8290	80
1976	65784	12521	2965	127929	82917	10957	82
1977	58849	10946	2889	125304	84168	10024	86
1978	51913	12521	2564	122679	85930	10308	90
1979	61326	12088	3096	124280	87051	11249	94
1980	70738	12822	3497	144507	10122	11560	100
1981	80151	13556	3898	180525	12645	12442	110
1982	58069	13587	3873	174147	13095	13931	117
1983	35986	15125	4279	178871	13765	15084	120
1984	55407	18465	4465	183595	14435	15687	126

Appendix 137

INDUSTRY: Chemical Fertilisers

Year	VALUE ADDED ('000)	WAGES & SALARIES ('000)	LABOR (NO OF PERSONS)	FIXED ASSETS ('000)	COST OF INPUT ('000)	DEPRE-CIATION ('000)	CPI (INDEX)
1963	2539	617	233	–	–	–	–
1964	4490	797	300	–	–	–	–
1965	4700	682	243	–	–	–	–
1966	4278	446	246	–	–	–	–
1967	5765	2061	681	–	–	–	–
1968	10605	2503	586	–	–	–	–
1969	15840	2647	609	35563	37682	7468	56
1970	16652	2965	712	32522	42018	6829	56
1971	12252	3004	690	31272	46288	6567	57
1972	15707	3146	671	28992	49126	6088	59
1973	36620	5201	1058	34412	80779	7227	65
1974	36292	7308	1340	33712	146407	7079	77
1975	54359	7494	1229	35472	138528	8290	80
1976	51623	9056	1371	43819	130748	10957	82
1977	58782	11286	1516	47115	155613	10504	86
1978	65940	13516	1661	50411	180477	10308	90
1979	93545	14584	1871	48923	282982	11249	94
1980	83428	17031	2063	56587	332139	11883	100
1981	73311	19477	2255	64251	381301	13492	110
1982	75576	22136	2274	716363	88899	15043	117
1983	77841	22441	1956	91032	396496	15084	120
1984	59733	23460	2028	98572	400157	20700	126

INDUSTRY: Paints, Varnish and Lacquer Industries

Year	VALUE ADDED ('000)	WAGES & SALARIES ('000)	LABOR (NO OF PERSONS)	FIXED ASSETS ('000)	COST OF INPUT ('000)	DEPRE- CIATION ('000)	CPI (INDEX)
1963	4647	1065	352	–	–	–	–
1964	5329	1202	394	–	–	–	–
1965	5824	1501	422	–	–	–	–
1966	7545	1751	475	–	–	–	–
1967	8550	1945	483	–	–	–	–
1968	10223	2346	546	–	–	–	–
1969	10001	2446	622	7677	15646	614	56
1970	10910	2694	655	7442	17844	595	56
1971	11903	3304	717	6696	20252	535	57
1972	14787	3577	717	7266	22900	581	59
1973	17279	4065	809	7578	30794	606	65
1974	19467	4893	874	10477	43947	838	77
1975	25110	5521	1004	11726	42768	948	80
1976	26854	6572	1021	10708	53640	840	82
1977	31954	7560	11111	3915	62047	1013	86
1978	37053	8547	1200	17122	70454	1020	90
1979	46220	10680	1365	22072	91984	1253	94
1980	53563	12911	1437	30497	107823	1829	100
1981	60905	15141	1508	38921	123661	2335	110
1982	67294	13332	1416	50419	127077	2520	117
1983	73683	17588	1384	57358	130493	2577	120
1984	75295	19027	1406	51075	130375	2553	126

INDUSTRY: Medicinal and Pharmacueticals

Year	VALUE ADDED ('000)	WAGES & SALARIES ('000)	LABOR (NO OF PERSONS)	FIXED ASSETS ('000)	COST OF INPUT ('000)	DEPRE-CIATION ('000)	CPI (INDEX)
1963	3897	526	419	–	–	–	–
1964	5068	595	486	–	–	–	–
1965	5655	700	491	–	–	–	–
1966	5218	822	534	–	–	–	–
1967	6555	888	588	–	–	–	–
1968	6758	1251	694	–	–	–	–
1969	4775	1580	629	1984	5680	119	56
1970	8270	1909	863	2970	7719	178	56
1971	6929	2128	953	3236	8667	194	57
1972	9799	2639	1151	5194	9940	311	65
1973	8699	3096	1365	4839	14205	290	65
1974	10018	3666	1418	6266	16424	375	77
1975	11960	4325	1199	8571	17956	526	80
1976	10287	5619	1394	9105	17004	726	82
1977	14379	7399	1579	9802	19271	735	86
1978	18470	9179	1764	10499	21538	913	90
1979	32447	10599	2196	16802	36968	1404	94
1980	36663	12459	2403	21649	39815	1731	100
1981	40879	14319	2609	26496	42662	2119	110
1982	42109	15528	2346	26413	48152	2113	117
1983	43338	14009	1943	26385	53642	2289	120
1984	50470	15691	1813	26996	51539	2294	126

INDUSTRY: Soaps & Detergents

Year	VALUE ADDED ('000)	WAGES & SALARIES ('000)	LABOR (NO OF PERSONS)	FIXED ASSETS ('000)	COST OF INPUT ('000)	DEPRE-CIATION ('000)	CPI (INDEX)
1963	13432	3229	1043	–	–	–	–
1964	20495	3981	1126	–	–	–	–
1965	16868	4363	1094	–	–	–	–
1966	20011	4234	1045	–	–	–	–
1967	17663	4236	953	–	–	–	–
1968	20026	4762	1069	–	–	–	–
1969	22361	4705	1052	8107	25394	567	56
1970	26023	4853	972	8878	26697	621	56
1971	24909	5298	958	9822	25998	687	57
1972	34337	5516	938	9463	28790	690	59
1973	50015	7157	1108	17980	35456	1258	65
1974	41111	7662	1097	19796	43617	1385	77
1975	28457	7301	1436	28565	56861	1907	80
1976	43248	7737	1215	28061	75177	2366	82
1977	54994	9014	1333	33232	75360	2658	86
1978	66740	10290	1451	38402	75542	3170	90
1979	91495	14997	1700	45029	81446	4013	94
1980	88297	15360	1621	49372	97294	4196	100
1981	85098	15722	1541	53715	93121	4565	110
1982	91567	17934	1580	60056	102193	4804	117
1983	98036	19407	1415	44946	111264	3332	120
1984	86351	22240	1392	53507	105672	3966	126

INDUSTRY: Perfumes, Cosmetics & Toiletteries

Year	VALUE ADDED ('000)	WAGES & SALARIES ('000)	LABOR (NO OF PERSONS)	FIXED ASSETS ('000)	COST OF INPUT ('000)	DEPRECIATION ('000)	CPI (INDEX)
1963	8755	704	369	–	–	–	–
1964	10896	1073	411	–	–	–	–
1965	11235	1049	430	–	–	–	–
1966	8641	1638	433	–	–	–	–
1967	9310	1755	485	–	–	–	–
1968	10628	1892	523	–	–	–	–
1969	12845	2344	724	5498	11835	329	56
1970	16217	2260	632	6392	19852	383	56
1971	18572	2697	675	6176	12204	370	57
1972	21097	2978	753	5842	17730	350	59
1973	19283	3083	868	5009	17206	301	65
1974	22786	3896	841	6709	22920	402	77
1975	5691	1652	567	3515	8596	215	80
1976	7122	1734	567	3656	9490	260	82
1977	7307	2145	661	3860	11865	270	86
1978	7491	2555	755	4664	14239	404	90
1979	15557	3706	699	11819	21428	899	94
1980	21556	5723	838	20320	34074	1016	100
1981	27554	7740	977	28821	46720	1046	110
1982	31663	10942	1012	28901	41679	1190	117
1983	35771	6945	608	21037	36638	1189	120
1984	36294	7851	704	22770	35831	1286	126

INDUSTRY: Petroleum Refineries

Year	VALUE ADDED ('000)	WAGES & SALARIES ('000)	LABOR (NO OF PERSONS)	FIXED ASSETS ('000)	COST OF INPUT ('000)	DEPRE-CIATION ('000)	CPI (INDEX)
1963	–	–	–	–	–	–	–
1964	–	–	–	–	–	–	–
1965	–	–	–	–	–	–	–
1966	–	–	–	–	–	–	–
1967	–	–	–	–	–	–	–
1968	40455	4457	378	–	–	–	–
1969	40205	4631	399	–	–	–	–
1970	42561	5033	422	–	–	–	–
1971	44917	5436	444	–	–	–	–
1972	47272	5838	467	–	–	–	–
1973	48628	6240	489	84174	182938	7575	65
1974	47600	7099	536	109345	580926	9841	77
1975	85513	7544	521	105761	679207	9368	80
1976	125021	7384	469	101581	881875	9740	82
1977	149814	8190	496	101588	881990	9142	86
1978	174607	8996	522	101595	882043	9358	90
1979	282791	9843	545	113270	983398	9267	94
1980	390975	12590	635	128796	1022075	11591	100
1981	499159	15336	725	161265	1060751	14513	110
1982	373884	19611	764	233896	1239162	21050	117
1983	248609	21931	917	519622	1417572	35921	120
1984	232548	27081	1395	533880	1491779	37371	126

INDUSTRY: Manufacture of Petroleum and Coal

Year	VALUE ADDED ('000)	WAGES & SALARIES ('000)	LABOR (NO OF PERSONS)	FIXED ASSETS ('000)	COST OF INPUT ('000)	DEPRE-CIATION ('000)	CPI (INDEX)
1963	–	–	–	–	–	–	–
1964	–	–	–	–	–	–	–
1965	–	–	–	–	–	–	–
1966	–	–	–	–	–	–	–
1967	–	–	–	–	–	–	–
1968	–	–	–	–	–	–	–
1969	–	–	–	–	–	–	–
1970	41878	4822	460	–	–	–	–
1971	44269	5406	556	147	155	8	57
1972	6724	6072	583	739	159	42	59
1973	2248	313	105	777	2382	45	65
1974	2724	384	144	1404	4235	81	77
1975	1203	558	170	1913	4997	112	80
1976	1430	482	102	1570	2540	89	82
1977	1633	513	106	1780	2880	103	86
1978	1835	544	106	1990	3220	119	90
1979	3165	822	198	2327	3765	169	94
1980	4496	1069	197	3151	8660	220	100
1981	5826	1315	195	3975	10925	278	110
1982	13443	2335	288	14264	21864	1070	117
1983	21059	4834	654	16517	32802	1243	120
1984	10805	3940	502	31652	47784	1382	126

INDUSTRY: Rubber Products Manufacturing

Year	VALUE ADDED ('000)	WAGES & SALARIES ('000)	LABOR (NO OF PERSONS)	FIXED ASSETS ('000)	COST OF INPUT ('000)	DEPRE-CIATION ('000)	CPI (INDEX)
1963	14088	7072	2253	–	–	–	–
1964	14865	7975	2658	–	–	–	–
1965	17097	11778	3158	–	–	–	–
1966	17845	9284	6071	–	–	–	–
1967	19717	9840	7248	–	–	–	–
1968	46618	16612	7715	–	–	–	–
1969	45992	17443	8390	42684	64586	1493	56
1970	51276	19496	8471	53868	68247	1885	56
1971	63315	19774	8835	47429	67702	1660	57
1972	72220	21922	9168	47463	81844	1661	59
1973	222451	52059	23995	143099	926925	5008	65
1974	211370	52059	22695	149123	932240	5219	77
1975	200290	52059	21395	155147	937555	5707	80
1976	189209	52059	20095	161171	942870	7498	82
1977	178128	52059	18795	167195	948185	8359	86
1978	167047	52059	17495	173219	953500	10226	90
1979	155967	52059	16195	179243	958815	10525	94
1980	144886	52059	14895	185267	964130	11116	100
1981	133805	52095	13595	197313	974763	15785	110
1982	426211	120651	21274	188775	920067	27949	117
1983	718616	189206	29853	180236	905371	26221	120
1984							

Appendix 145

INDUSTRY: Manufacture of Plastic Products

Year	VALUE ADDED ('000)	WAGES & SALARIES ('000)	LABOR (NO OF PERSONS)	FIXED ASSETS ('000)	COST OF INPUT ('000)	DEPRE-CIATION ('000)	CPI (INDEX)
1964	–	–	–	–	–	–	–
1965	–	–	–	–	–	–	–
1966	–	–	–	–	–	–	–
1967	–	–	–	–	–	–	–
1968	–	–	–	–	–	–	–
1969	–	–	–	–	–	–	–
1970	–	–	–	–	–	–	–
1971	18580	7647	5091	24586	33200	2212	57
1972	27822	8890	5934	34530	47955	3107	59
1973	44488	12689	8614	48832	88224	4394	65
1974	48570	15606	7907	62064	93368	5585	77
1975	38403	15642	7120	66840	82201	6513	80
1976	48119	20596	8151	93027	133539	8599	82
1977	71990	25233	9792	103896	149141	9350	86
1978	95860	29869	11432	114764	164741	13007	90
1979	121691	41783	13733	149822	215093	16378	94
1980	147521	53696	16035	202409	293783	18216	100
1981	17335	65610	18336	254996	372474	22949	110
1982	181106	62448	14747	257513	369165	23176	117
1983	188859	66976	14250	266155	365856	23908	120
1984	179270	87130	14655	300134	376681	26961	126

INDUSTRY: Manufacture of Pottery, China and Eathernware

Year	VALUE ADDED ('000)	WAGES & SALARIES ('000)	LABOR (NO OF PERSONS)	FIXED ASSETS ('000)	COST OF INPUT ('000)	DEPRE- CIATION ('000)	CPI (INDEX)
1963	868	404	283	–	–	–	–
1964	830	412	250	–	–	–	–
1965	812	370	268	–	–	–	–
1966	1031	538	344	–	–	–	–
1967	1310	575	314	–	–	–	–
1968	1139	547	280	–	–	–	–
1969	1594	645	357	2324	1341	92	56
1970	1495	541	276	2424	909	97	56
1971	1601	513	253	2393	934	96	57
1972	3368	1216	533	4893	2077	195	59
1973	6294	1973	946	6575	4669	263	65
1974	8148	1988	914	21006	5448	840	77
1975	7572	2174	584	21601	5336	885	80
1976	7562	2884	1074	22707	6763	870	82
1977	10136	3967	1450	26156	10634	1046	86
1978	12710	5050	1825	29604	14504	1887	90
1979	15793	5466	1741	31677	14563	1989	94
1980	21256	7292	1947	37543	17305	2252	100
1981	26718	9118	2153	43408	20046	2604	110
1982	251404	9123	2024	50169	22090	3010	117
1983	28650	9128	1792	56930	24133	3273	120
1984	25057	10256	1914	86279	123840	4831	126

INDUSTRY: Hydraulic Cement Manufacturing

Year	VALUE ADDED ('000)	WAGES & SALARIES ('000)	LABOR (NO OF PERSONS)	FIXED ASSETS ('000)	COST OF INPUT ('000)	DEPRE- CIATION ('000)	CPI (INDEX)
1963	14071	2314	1036	–	–	–	–
1964	14071	2314	1036	–	–	–	–
1965	23030	3083	1027	–	–	–	–
1966	15907	3691	1159	–	–	–	–
1967	32049	3569	1109	–	–	–	–
1968	25485	4323	1113	–	–	–	–
1969	40136	4929	1145	56662	28662	3399	56
1970	45178	4981	1181	52487	28399	3149	56
1971	44869	5572	1189	49358	32068	2961	57
1972	55855	5876	1159	45968	39255	2758	59
1973	44221	6527	1278	66922	42116	4015	65
1974	50268	7696	1356	110380	69117	6622	77
1975	56100	9881	1657	141967	77095	7319	80
1976	58513	10872	1564	174743	97215	9074	82
1977	68473	13806	1827	161211	106981	11284	86
1978	78433	16739	2089	187678	116747	14031	90
1979	104076	18626	2299	188590	140872	14303	94
1980	124488	23833	2709	339220	123260	23745	100
1981	144900	29039	3119	489850	150648	29391	110
1982	207685	35337	3049	493622	235053	29617	117
1983	270469	40706	2988	543259	364457	30105	120
1984	346423	43638	3016	729346	314023	43760	126

INDUSTRY: Cement and Concrete

Year	VALUE ADDED ('000)	WAGES & SALARIES ('000)	LABOR (NO OF PERSONS)	FIXED ASSETS ('000)	COST OF INPUT ('000)	DEPRE-CIATION ('000)	CPI (INDEX)
1963	6407	2440	1339	–	–	–	–
1964	11129	3337	1769	–	–	–	–
1965	10385	3599	1677	–	–	–	–
1966	12467	3852	1601	–	–	–	–
1967	10542	4951	1806	–	–	–	–
1968	40136	5056	2090	–	–	–	–
1969	16778	5023	1989	20771	21800	1370	56
1970	20347	5546	2165	20090	21620	1326	56
1971	27957	5416	2320	21749	23549	1435	57
1972	27957	5416	2320	24744	28251	1633	59
1973	32680	8927	3112	29849	37992	1970	65
1974	40902	10335	3347	39563	57277	2611	77
1975	19588	12452	3385	47984	57277	3199	80
1976	20976	13107	3371	73864	78468	5407	82
1977	47812	15787	3770	73548	87766	6170	86
1978	74647	18467	4169	73232	97063	6160	90
1979	96561	25613	5339	102409	130660	8759	94
1980	112000	27326	6797	138923	188935	11808	100
1981	144900	29039	8254	175437	247310	14912	110
1982	207685	35337	6742	176269	249976	14982	117
1983	270469	40706	6570	289551	252642	29607	120
1984	202900	58493	6703	329242	289242	27986	126

INDUSTRY: Primary Iron and Steel Industry

Year	VALUE ADDED ('000)	WAGES & SALARIES ('000)	LABOR (NO OF PERSONS)	FIXED ASSETS ('000)	COST OF INPUT ('000)	DEPRE-CIATION ('000)	CPI (INDEX)
1963	–	–	–	–	–	–	–
1964	–	–	–	–	–	–	–
1965	–	–	–	–	–	–	–
966	–	–	–	–	–	–	–
1967	–	–	–	–	–	–	–
1968	18111	6562	2416	–	–	–	–
1969	26785	6476	2106	–	–	–	–
1970	28179	7592	2411	82305	70482	9053	56
1971	32687	8115	2360	82956	71040	9125	57
1972	42518	8115	2829	96974	89246	9697	59
1973	39332	9685	2207	75583	57629	8314	65
1974	44849	8313	2217	69053	54058	7595	77
1975	57936	8738	2269	87756	74435	9766	80
1976	39711	11424	2559	82545	75507	9587	82
1977	47061	14945	2374	70451	53380	9463	86
1978	54410	14995	2189	58356	62544	8504	90
1979	51254	15044	2016	59300	65503	9526	94
1980	48097	16239	2095	69215	123341	8305	100
1981	44941	24699	2174	85130	181178	10215	110
1982	65354	23077	2103	99298	244981	11915	117
1983	85767	31447	2691	183959	308783	12284	120
1984	97969	32617	3345	189819	414226	12676	126

INDUSTRY: Non–Ferous Metal Industries

Year	VALUE ADDED ('000)	WAGES & SALARIES ('000)	LABOR (NO OF PERSONS)	FIXED ASSETS ('000)	COST OF INPUT ('000)	DEPRE-CIATION ('000)	CPI (INDEX)
1963	1751	868	377	–	–	–	–
1964	5474	1515	586	–	–	–	–
1965	5025	1567	623	–	–	–	–
1966	6821	1677	606	–	–	–	–
1967	8043	1764	604	–	–	–	–
1968	2036	1814	239	–	–	–	–
1969	4762	1024	307	6407	5303	480	56
1970	3374	1029	320	7764	5820	582	56
1971	4273	1972	402	9066	6825	679	57
1972	4296	1965	563	9394	12275	704	59
1973	9742	2530	686	9705	11424	727	65
1974	11687	3182	761	16320	21766	1224	77
1975	7468	4299	780	17093	22941	1320	80
1976	10296	5429	967	19544	28961	1478	82
1977	13524	6310	1065	20175	36763	1513	86
1978	16751	7191	1162	20805	44565	1712	90
1979	26149	9141	1372	34485	59268	2347	94
1980	26069	11028	1534	42309	77118	3173	100
1981	25989	12915	1695	50132	94967	3759	110
1982	67935	27325	3053	172632	137057	8631	117
1983	109881	31455	3248	188039	140567	8381	120
1984	109244	32855	3148	203445	144076	10172	126

INDUSTRY: Wire Products Manufacturing

Year	VALUE ADDED ('000)	WAGES & SALARIES ('000)	LABOR (NO OF PERSONS)	FIXED ASSETS ('000)	COST OF INPUT ('000)	DEPRE-CIATION ('000)	CPI (INDEX)
1963	1684	627	460	–	–	–	–
1964	1401	719	487	–	–	–	–
1965	1985	716	510	–	–	–	–
1966	2287	750	509	–	–	–	–
1967	2650	844	534	–	–	–	–
1968	3376	1041	646	–	–	–	–
1969	7504	1188	797	3521	13184	281	56
1970	5298	1782	1041	6485	16717	518	56
1971	8018	2796	1522	7467	21230	597	57
1972	9461	3134	1846	24592	32398	1967	59
1973	23970	3628	2167	26010	52144	2080	65
1974	27870	4865	1885	34661	61679	2772	77
1975	20465	4927	1774	34216	54715	3024	80
1976	22182	6150	2003	38509	67220	3159	82
1977	32512	7483	2240	44631	94996	3570	86
1978	42841	8815	2477	50753	112772	3530	90
1979	61509	11484	2863	58845	164270	4464	94
1980	55032	15034	3229	65723	165444	5257	100
1981	48555	18584	3594	72601	166618	5808	110
1982	57704	17194	2699	92661	199852	7412	117
1983	66853	20344	2742	125385	233086	7525	120
1984	49315	22686	2806	116009	186852	6960	126

INDUSTRY: Brass, Copper, Pewter & Alluminium Products

Year	VALUE ADDED ('000)	WAGES & SALARIES ('000)	LABOR (NO OF PERSONS)	FIXED ASSETS ('000)	COST OF INPUT ('000)	DEPRE-CIATION ('000)	CPI (INDEX)
1963	2525	929	611	–	–	–	–
1964	2525	1029	611	–	–	–	–
1965	3205	1212	684	–	–	–	–
1966	3282	1298	947	–	–	–	–
1967	4012	1605	947	–	–	–	–
1968	5250	1971	1086	–	–	–	–
1969	6029	1985	1247	4978	8930	298	56
1970	5938	2223	1119	4974	9137	299	56
1971	5858	2668	1302	5868	9207	352	57
1972	8522	3021	1580	6702	10615	402	59
1973	12940	4581	2480	9103	18481	546	65
1974	13521	6661	2719	15601	24518	936	77
1975	18165	6528	2524	18567	23739	1274	80
1976	26148	9010	2671	20291	31946	1221	82
1977	27518	9302	2610	21673	38018	1300	86
1978	28888	9593	2548	23054	44090	1754	90
1979	32214	9533	2861	24879	53340	1652	94
1980	42869	15246	3677	46653	77255	2799	100
1981	53524	20958	4493	68427	101169	4105	110
1982	55361	22307	3946	79689	101668	4781	117
1983	67878	21037	3184	71751	102167	4104	120
1984	50616	23860	3212	60072	100350	3604	126

INDUSTRY: Industrial Machinery and Parts

Year	VALUE ADDED ('000)	WAGES & SALARIES ('000)	LABOR (NO OF PERSONS)	FIXED ASSETS ('000)	COST OF INPUT ('000)	DEPRE-CIATION ('000)	CPI (INDEX)
1963	9290	5286	774	–	–	–	–
1964	9628	5465	869	–	–	–	–
1965	12567	6797	921	–	–	–	–
1966	14293	7984	991	–	–	–	–
1967	15217	4054	1054	–	–	–	–
1968	15420	8086	1109	–	–	–	–
1969	17482	8712	1340	7143	19612	500	56
1970	17811	9405	1344	7058	23521	494	56
1971	19671	10416	1349	7645	25113	535	57
1972	23406	11428	1445	10797	28102	755	59
1973	8076	4072	1444	7248	12690	507	65
1974	11584	4713	1448	9122	16047	638	77
1975	15884	4494	1320	6017	11206	460	80
1976	9392	4006	1279	7828	16070	655	82
1977	11142	4283	1235	6784	17747	650	86
1978	12892	4559	1191	5739	17217	622	90
1979	13010	5049	170	6892	31318	737	94
1980	19918	7774	1614	11384	50577	1138	100
1981	26856	10498	2057	15875	69835	1587	110
1982	33905	13402	2090	17159	83032	1715	117
1983	40983	17305	2339	21861	96228	2235	120
1984	31459	15421	2071	28184	106349	2828	126

INDUSTRY: Manufacture of Electrical Machinery and Apparatus

Year	VALUE ADDED ('000)	WAGES & SALARIES ('000)	LABOR (NO OF PERSONS)	FIXED ASSETS ('000)	COST OF INPUT ('000)	DEPRE-CIATION ('000)	CPI (INDEX)
1963	–	–	–	–	–	–	–
1964	–	–	–	–	–	–	–
1965	–	–	–	–	–	–	–
1966	–	–	–	–	–	–	–
1967	–	–	–	–	–	–	–
1968	–	–	–	–	–	–	–
1969	–	–	–	–	–	–	–
1981	26856	10498	2057	15875	69835	1587	110
1970	32662	8489	3206	33941	36758	4072	56
1971	26569	11869	3787	65357	70781	7842	57
1972	72421	15248	3869	54185	105638	8502	59
1973	188527	41462	25332	187307	222917	22476	65
1974	259143	65782	26669	270458	586200	24750	77
1975	334025	94014	33145	212005	654843	26005	80
1976	516752	129730	45550	280996	963768	34381	82
1977	494457	163299	53011	338147	1159789	40577	86
1978	572162	196868	60472	395298	1355805	51879	90
1979	792988	264850	72770	503061	1725411	71495	94
1980	1013814	327569	76944	717923	2462161	86150	100
1981	1234640	390287	81118	932425	3131673	111891	110
1982	1368140	460291	78372	1113984	3634904	155957	117
1983	1612408	547122	86861	1241831	4138134	193285	120
1984	2019015	638511	92980	1640294	4864840	209641	126

INDUSTRY: Shipbuilding, Boatmaking and Repair Services

Year	VALUE ADDED ('000)	WAGES & SALARIES ('000)	LABOR (NO OF PERSONS)	FIXED ASSETS ('000)	COST OF INPUT ('000)	DEPRE- CIATION ('000)	CPI (INDEX)
1963	3458	2029	837	–	–	–	–
1964	2712	1169	822	–	–	–	–
1965	3516	2316	879	–	–	–	–
1966	3123	1524	792	–	–	–	–
1967	1988	1338	574	–	–	–	–
1968	1955	1436	467	–	–	–	–
1969	2418	1385	542	4883	2185	195	56
1970	2562	1603	511	5062	1706	202	56
1971	3041	2795	551	4995	2363	199	57
1972	4718	3261	923	11302	2731	452	59
1973	7004	4412	1113	15719	5256	628	65
1974	12880	5055	1475	18512	12018	740	77
1975	40905	9163	1558	19982	22747	1293	80
1976	25125	13361	1966	186618	45720	2472	82
1977	26311	17558	2032	187320	46852	7092	86
1978	27496	18347	2098	188022	46852	7244	90
1979	74117	27921	2282	175919	24559	6508	94
1980	125954	37494	4245	195533	103500	7821	100
1981	177790	56640	6208	215267	182441	8610	110
1982	126468	44392	4086	197910	139382	8466	117
1983	75145	29783	3313	195964	136322	8384	120
1984	108302	38315	2968	197323	163480	8892	126

INDUSTRY: Motor Vehicle Bodies

Year	VALUE ADDED ('000)	WAGES & SALARIES ('000)	LABOR (NO OF PERSONS)	FIXED ASSETS ('000)	COST OF INPUT ('000)	DEPRE-CIATION ('000)	CPI (INDEX)
1963	1660	1084	497	–	–	–	–
1964	1954	1214	519	–	–	–	–
1965	1842	1171	573	–	–	–	–
1966	1998	1081	545	–	–	–	–
1967	1721	1028	469	–	–	–	–
1968	1819	1093	488	–	–	–	–
1969	1842	1222	540	540	2939	37	56
1970	2255	1344	599	777	5656	54	56
1971	3143	1799	738	969	8296	67	57
1972	3012	1662	592	1421	7217	99	59
1973	4472	2053	823	1668	9912	116	65
1974	5201	2440	759	1765	10381	123	77
1975	4576	2511	801	1891	10800	134	80
1976	7123	3981	921	2853	13822	312	82
1977	9689	4506	1048	3154	19567	346	86
1978	12255	5030	1175	3455	25312	412	90
1979	17703	6278	1498	4883	35505	599	94
1980	21669	8934	1706	8377	42385	921	100
1981	25634	11589	1913	8462	49265	930	110
1982	26153	11288	1582	8547	48860	940	117
1983	26672	12874	1699	10723	48454	1098	120
1984	30243	15543	1691	21848	56684	2403	126

INDUSTRY: Motor Vehicle Parts and Accessories

Year	VALUE ADDED ('000)	WAGES & SALARIES ('000)	LABOR (NO OF PERSONS)	FIXED ASSETS ('000)	COST OF INPUT ('000)	DEPRE-CIATION ('000)	CPI (INDEX)
1963	227	112	85	–	–	–	–
1964	332	153	97	–	–	–	–
1965	416	164	102	–	–	–	–
1966	366	172	100	–	–	–	–
1967	473	192	126	–	–	–	–
1968	547	230	141	–	–	–	–
1969	727	303	214	588	1946	35	56
1970	797	401	255	942	1586	56	56
1971	789	403	261	919	1181	55	57
1972	890	585	336	2819	1535	169	59
1973	1596	733	404	2838	2211	170	65
1974	3997	1309	602	8406	4575	504	77
1975	6292	2749	1133	15307	8428	982	80
1976	8483	3531	1452	18032	14877	1367	82
1977	12005	4101	1634	21555	17582	1616	86
1978	15526	4670	1815	25077	20189	2068	90
1979	20603	6933	2146	35333	25825	3136	94
1980	30010	11200	2966	60263	42678	4821	100
1981	39417	15467	3373	58192	59531	6815	117
1982	54445	16446	3421	69664	68914	5573	117
1983	69472	19541	3490	84647	78296	8151	120
1984	74340	23567	3575	90882	83776	8179	126

INDUSTRY: Manufacture and Assembly of Bicycles, Parts and Accessories

Year	VALUE ADDED ('000)	WAGES & SALARIES ('000)	LABOR (NO OF PERSONS)	FIXED ASSETS ('000)	COST OF INPUT ('000)	DEPRE-CIATION ('000)	CPI (INDEX)
1963	238	75	85	–	–	–	–
1964	261	77	62	–	–	–	–
1965	303	72	60	–	–	–	–
1966	338	88	84	–	–	–	–
1967	581	154	160	–	–	–	–
1968	621	430	255	–	–	–	–
1969	3297	839	336	3644	5926	236	56
1970	2701	872	346	3473	4215	225	56
1971	2353	1080	456	4015	4432	260	57
1972	3884	1189	548	4396	8913	285	59
1973	5811	1781	899	6901	12794	448	65
1974	6604	2206	827	8130	14126	528	77
1975	4118	2616	1012	9172	11099	613	80
1976	5520	3137	1135	10366	14438	650	82
1977	6062	3341	1122	12001	16738	780	86
1978	6603	3544	1110	16360	19037	804	90
1979	10597	4677	1270	19642	23821	756	94
1980	11700	5578	1366	20542	29954	821	100
1981	12803	6479	1462	29441	36086	1177	110
1982	12292	5925	1184	23787	31953	1137	117
1983	11781	5867	1265	27178	27819	1165	120
1984	11340	6192	1201	28016	33745	1204	126

INDUSTRY: Manufacture of Professional and Scientific Equipment

Year	VALUE ADDED ('000)	WAGES & SALARIES ('000)	LABOR (NO OF PERSONS)	FIXED ASSETS ('000)	COST OF INPUT ('000)	DEPRE-CIATION ('000)	CPI (INDEX)
1963	–	–	–	–	–	–	–
1964	–	–	–	–	–	–	–
1965	–	–	–	–	–	–	–
1966	–	–	–	–	–	–	–
1967	–	–	–	–	–	–	–
1968	–	–	–	–	–	–	–
1969	–	–	–	–	–	–	–
1970	4221	1421	423	2676	4542	161	56
1971	4856	1781	666	4375	7425	262	57
1972	10864	2360	958	6568	10864	294	59
1973	9925	2950	984	7030	13178	421	65
1974	9510	3144	906	8285	13025	497	77
1975	7268	3835	948	9793	13451	561	80
1976	19335	9838	3144	9951	13961	744	82
1977	28120	12120	3144	22337	31338	893	86
1978	36904	14401	3144	34722	48714	928	90
1979	32605	18074	4024	38825	54471	1355	94
1980	28307	23126	4429	45955	66799	1378	100
1981	24008	28177	4934	58085	79127	1742	110
1982	50385	2419	4239	56224	76949	2248	117
1983	76761	30150	5604	55480	85877	2294	120
1984	59112	26241	4843	51700	88620	2137	126